THE CET STUDY GUIDE

THE CET STUDY GUIDE

BY SAM WILSON, CET
ISCET TEST CONSULTANT

TAB

TAB BOOKS Inc.
Blue Ridge Summit, PA 17214

FIRST EDITION

THIRD PRINTING

Printed in the United States of America

Library of Congress Cataloging in Publication Data

Wilson, J. A.
The CET study guide.

Includes index.
1. Electronics—Examinations, questions, etc.
2. Electronic technicians—Certification—United States.
I. Title. II. Title: C.E.T. study guide.
TK7863.W53 1984 621.381'076 84-8517
ISBN 0-8306-0791-9
ISBN 0-8306-1791-4 (pbk.)

Contents

Introduction ix

1 **What You Should Know About the CET Test** 1
The CET Test—Important Rules for Taking the CET Test—Programmed Review

2 **Two-Terminal Components** 15
Linear, Bilateral Circuit Elements—Nonlinear Two-Terminal Components—Programmed Review—Additional Practice CET Test Questions—Answers to Practice Test

3 **Three-Terminal Amplifying Components and Basic Circuits** 44
Basic Electric Circuits—Three-Terminal Amplifying Components—Noise in Amplifiers—Distortion in Amplifiers—Special Ratings—Programmed Review—Additional Practice CET Test Questions—Answers to Practice Test

4 **Antennas and Transmission Lines** 80
Transmission Lines—Installation of Transmission Lines—Antennas—Antenna Installations—Programmed Review—Additional Practice CET Test Questions—Answers to Practice Test

5 **Digital Circuits in Consumer Products** 105
Gate Symbols and Other Identification—Important Rules for Overbars—Counting Systems—Some Basic Facts on Hardware—Flip-Flops—Memories and Microprocessors—An Introduction to Microprocessor Terminology—Other Important Integrated Circuit Logic Systems and

Devices—Additional Terms You Should Know for the CET Test—Programmed Review—Additional Practice CET Test Questions—Answers to Practice Test

6 Linear Circuits in Consumer Products 151
Integrated Circuit Operational Amplifiers—Phase-Locked Loop—Additional Analog Circuits—Amplifier Coupling Circuits—Power Supply Circuits—Programmed Review—Additional Practice CET Test Questions—Answers to Practice Test

7 Television 188
Television and FM Signals—Color TV—FM—The Television Receiver—Video Recorders and the CET Test—Programmed Review—Additional Practice CET Test Questions—Answers to Practice Test

8 Test Equipment and Troubleshooting 220
Basic Meter Movements—Oscilloscopes—Evaluating Parameters—Testing Amplifiers—Programmed Review—Additional Practice CET Test Questions—Answers to Practice Test

Appendix A Sample Journeyman CET Test 257
Section 9, Antennas and Transmission Lines—Section 10, Digital Circuits in Consumer Products—Section 11, Linear Circuits in Consumer Products—Section 12, Television—Section 13, Videocassette Recorders—Section 14, Troubleshooting Consumer Equipment—Section 15, Test Equipment—Answers to Practice CET Test

Index 279

THE CET STUDY GUIDE

Introduction

Being a certified electronics technician (CET) wouldn't mean much if you could get the certificate just for the asking. To become a CET it is necessary to pass a 75-question Associate-level CET Test. Passing that test is required of all CETs. A 75-question Journeyman-level CET Test in the specialty of the technician must also be passed.

By the time most technicians get ready to take their Journeyman-level CET Test, they've been out of school for at least four years. What they need is a study guide that allows them to touch up on their theory, identify and re-study those subjects which have faded over the years, and brush up on some of the new technology. That is the purpose of this book—to provide a quick, easy reference so that experienced technicians don't have to go into the test "cold turkey."

This is not a textbook. Its purpose is to fill in the gaps and provide an overall study review. If there are subjects in this book that are unfamiliar to you, you should get one of the many excellent textbooks on the market and review those subjects in greater depth. If you are a journeyman technician, you have surely been taking the technical magazines. Consult the table of contents in those magazines to find review material for bringing your technology up to date.

In some cases, you may find questions in this book that have answers that you don't agree with. Please feel free to write to me personally c/o TAB BOOKS Inc., and I will gladly discuss these subjects with you. That will help me as much as it will help you, because sooner or later this book will be revised. I always give those letters serious consideration when I update a book. When you get ready to take the test, send a note to ISCET (International Society of Certified Electronics Technicians) at the following address:

ISCET
2708 West Berry St.
Fort Worth, TX 76109

They will provide a list of certification administrators in your area.

I never say "good luck" to someone who is going to take a CET test. I think that's an insult because there's no luck involved. You are a highly skilled technician and you know that passing the CET test is a matter of knowing your subject. That can only come with study. And I would like to say one additional thing. *If you don't pass the first time, it doesn't mean that you're not a good technician.* It means that there are one or two technical subjects that you should review. After all, you know the importance in keeping your technology up to date. So, if that happens go back to the study guide. Review with increased intensity. Then take the test again. When you can say "I'm a CET," I am certain you will feel that it was well worth the effort.

Feel free to write to me about the subjects in this book or about your CET test results. A self-addressed envelope would be appreciated.

Sam Wilson—CET ISCET Test Consultant
1943 East Indian School Road
Phoenix, AZ 85016

What You Should Know About the CET Test

In this chapter, you will be learning about the CET test that you will be taking for your Journeyman CET rating. Some helpful hints will be given about how to take the test, what material is in the test, and ISCET's rules related to taking the test. Figure 1-1 shows a Journeyman CET certificate.

THE CET TEST

The first step in becoming a Journeyman-level CET is to take the Associate-level Test. Many of the readers of this book will have already taken the Associate-level Test, but it is not uncommon for a technician to take the Associate-level and Journeyman-level Tests together, or at least, within a few days of each other. So, it is helpful to review some of the material you will encounter in the Associate-level Test.

Here is a list of the headings of each section of the Associate-level Test, and the number of questions asked under each heading:

Section I Basic Mathematics	5 questions
Section II Dc Circuits	5 questions
Section III Ac Circuits	5 questions

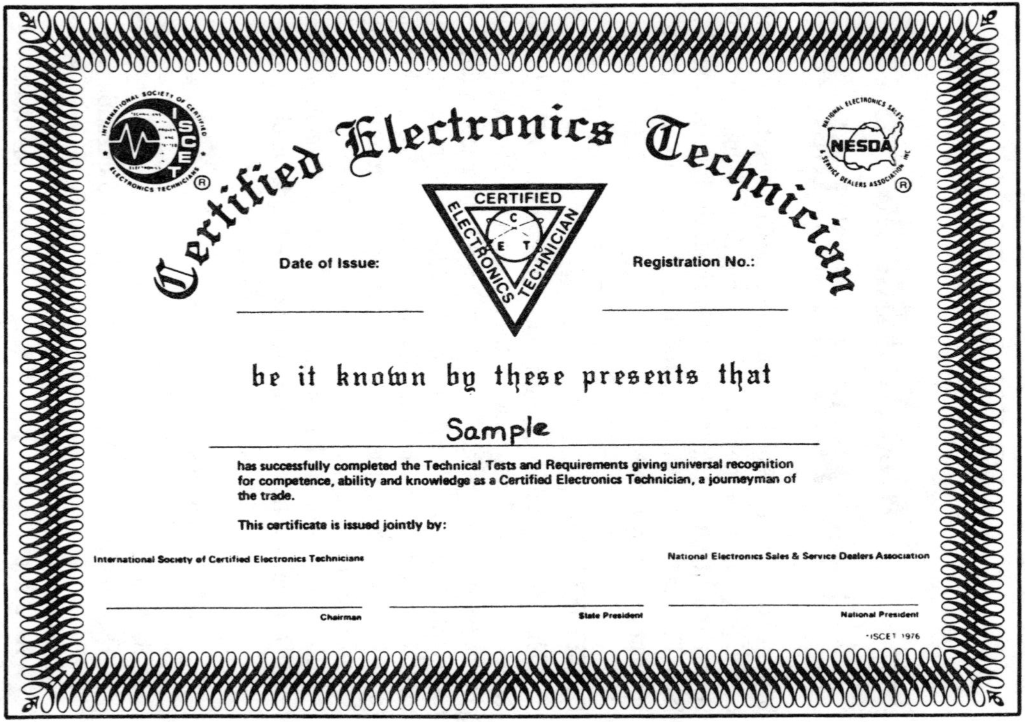

Certified Electronics Technician

Date of Issue: ____________

Registration No.: ____________

be it known by these presents that

Sample

has successfully completed the Technical Tests and Requirements giving universal recognition for competence, ability and knowledge as a Certified Electronics Technician, a journeyman of the trade.

This certificate is issued jointly by:

International Society of Certified Electronics Technicians

National Electronics Sales & Service Dealers Association

Chairman

State President

National President

©ISCET 1976

Fig. 1-1. Journeyman-level CET certificate.

Section IV Transistors and Semiconductors	10 questions
Section V Electronic Components	10 questions
Section VI Instruments	10 questions
Section VII Tests and Measurements	15 questions
Section VIII Troubleshooting and Network Analysis	15 questions

Notice that the Mathematics section has only five questions. This was the most controversial part of the Associate-level Test when it first was written. Many technicians claim that they have years of experience without ever using any mathematics. This isn't possible. Actually, they use mathematics every time they make a measurement; and, measurement is an important part of the troubleshooting procedure.

Here is an example: A technician makes a measurement of voltage across a resistor. That voltage reading is too low according to the manufacturer's callout on the schematic diagram. What does a technician know immediately? Well, he knows that either *the current is too low* through the resistor, or, the resistance has changed to a *lower value*. Those are the two immediate possibilities.

How does he know these things? Because he understands Ohm's law, and he knows that the *voltage is directly related to the amount of current through a resistor and the resistance of that resistor.* He knows that relationship because he worked many problems in Ohm's law when he was learning basic electronics.

The most important thing about the mathematics for a technician is that it gives him an understanding of how the different parameters in the circuit are related. It is highly doubtful that a technician could work efficiently without having done such problems. If he had to learn the relationship between voltage, current, and resistance in a circuit by words rather than by the simple Ohm's law equation, it would have taken much longer to learn how voltage, current, and resistance are related. Suppose, for example, instead of V = IR he had to learn the following expression: "Whenever there is a voltage across a resistor, the amount of that voltage is directly proportional to the current through the resistor, and also directly proportional to the resistance value of the resistor."

That statement would only explain the relationship of current and resistance voltage. Another similar statement would be required for current, and another one would be needed for resistance. Can you image trying to learn the relationships between current, voltage, and resistance that way?

Don't be put off by the mathematics section in the Associate-level Test. And also, don't be put off by the very small amount of mathematics that you will encounter in the Journeyman-level Test. It is not intended that this mathematics be of any importance other than how it shows the relationship between the circuit parameters.

After taking the Associate-level Test, which is required for all technicians, the next step is to take the Journeyman-level Test. There are a number of different options available, and the technician will naturally take that test that most closely relates to his four years of experience. Among the options available:

- Consumer electronics (radio, TV, and some audio)
- Audio
- Industrial electronics
- Communications
- Medical electronics
- MATV/CATV
- Computer technology
- Radar

Remember that you need four years of experience in the option that you select. It is possible to get some credit against that four years if you have attended an approved school. If you feel that rule applies to you, write to the ISCET office and ask them for futher details.

The major thrust of this book is toward the Journeyman-level Consumer CET Test. Here is an outline of that test along with the number of questions in each section:

Section IX Antennas and Transmission Lines	5 questions
Section X Digital Circuits in Consumer Products	15 questions

Section XI Linear Circuits in Consumer Products	15 questions
Section XII Television	15 questions
Section XIII Videocassette Recorders	5 questions
Section XIV Troubleshooting Consumer Equipment	15 questions
Section XV Test Equipment	5 questions
Total	75 questions

Remember that each CET Test (Associate-level and Journeyman-level) is a 75-question, multiple-choice test. That means you will be given a question and number of possible answers, and you are to select the correct one in each case.

Be very careful when you are filling in your answer sheet. There are two types in use as shown in Fig. 1-2. Technicians sometimes assume that the answer sheet is numbered horizontally, or, they assume it is vertically numbered. So, they make a mistake when filling in the answers without checking. After they have completed a number of answers they begin to realize that the numbers don't match the questions they are answering and they have to start over. That puts them in a hurry, and *one sure way to fail a CET Test is to get in a hurry to answer the questions.*

There is another important point to remember about numbering: *When you are taking the Journeyman-level Test, the first question is numbered 76.* The reason for that, of course, is that numbers 1 through 75 are in the Associate-level Test. So, you do not start answering questions in the Journeyman-level Test at the top of the answer sheet. Instead, your first answer will go beside the space for answer number 76.

There are no trick questions in the CET test. Sometimes technicians carelessly answer a question and get it wrong. Or, due to the fact that they haven't thoroughly understood or studied some particular facet of electronics, they get a wrong answer.

ANSWER SHEET
PAGE 1

I.S.C.E.T CERTIFICATION TESTSN 263-35

NAME ____________________

DATE ____________________

LOCATION ____________________

PROCTOR ____________________

1	A B C D
2	A B C D
3	A B C D
4	A B
5	A B
6	A B C D
7	A B
8	A B C D
9	A B C
10	A B C D
11	A B C D
12	A B C D
13	A B C D
14	A B C D
15	A B C D
16	A B C D
17	A B C D
18	A B
19	A B C D
20	A B C D
21	A B C D
22	A B C D
23	A B
24	A B C D
25	A B C D
26	A B C D
27	A B C D
28	A B C D
29	A B
30	A B C D
31	A B C D
32	A B C D
33	A B C D

Fig. 1-2. Two types of answer sheets are used. Be sure to enter your answers correctly.

I S.C.E.T CERTIFICATION TESTSN 263-35

34. A B C D
35. A B C D
36. A B C D
37 A B C D
38. A B C D
39 A B
40 A B C D
41 A B C D
42. A B C D
43. A B C D
44. A B C D
45. A B C D
46. A B
47. A B C D
48. A B C D
49 A B C D
50. A B C D
51 A B
52. A B
53. A B C D
54. A B C
55. A B
56. A B C D
57 A B
58. A B C D
59 A B
60. A B C D
61. A B C D
62. A B C D
63. A B C D
64. A B
65. A B C D
66. A B C
67. A B C D
68. A B C D
69. A B C D
70. A B C D
71. A B C D
72. A B C
73. A B C D
74. A B
75. A B

NAME ___ LAST FIRST MIDDLE DATE ___ AGE ___ SEX ___ DATE OF BIRTH ___

SCHOOL ___ CITY ___ GRADE OR CLASS ___ INSTRUCTOR ___

NAME OF TEST ___ PART ___

DIRECTIONS: Read each question and its numbered answers. When you have decided which answer is correct, blacken the corresponding space on this sheet with a No. 2 pencil. Make your mark as long as the pair of lines, and completely fill the area between the pair of lines. If you change your mind, erase your first mark COMPLETELY. Make no stray marks; they may count against you.

SAMPLE

I. CHICAGO is
1-1 a country 1-4 a city
1-2 a mountain 1-5 a state
1-3 an island

PLEASE RETAIN THIS RECEIPT. This information is necessary if you inquire about your exam:

NAME ___ Exam Number ___ Date ___

Amount Paid $___ ___ Examiner's Signature

Fig. 1-2. Two types of answer sheets are used. Be sure to enter your answer correctly. (Continued from page 7.)

When they do miss a question, there is a tendency for them to think that it is a trick question; and, it is not uncommon for a technician to get angry if he believes he has been tricked on a CET test. Let me say this again, and it is very important: *there are no trick questions in the CET test!*

Every effort has been made to analyze each question to make sure it is fair, and that it is a question an experienced technician should be able to answer. There are committees that pass on these tests—they are *not* the work of one single person. These committees are made up of experienced technicians who are now CETs. They have a responsibility to protect the integrity of the CET test, and therefore, they don't approve giveaway questions. Also, they do not want questions that are unrelated to the technician's workday experience.

No specialized knowledge of any particular manufacturer's equipment is necessary in order to pass the CET test. You will not, for example, find a question on the startup circuit for an RCA scan-derived power supply. Of course, any technician who works for that manufacturer could quickly answer such a question, but technicians unfamiliar with that particular circuit would not have that specialized knowledge. So, the questions in the CET test are about general knowledge, and about applications that apply to all manufactured equipment.

You will be given plenty of time to take the test. It is not a time-limited test. The Test Administrators have been instructed to permit the examinees plenty of time. Of course, if you come to the point when you are no longer making any progress, it is useless to waste your time and the examiner's time by continuing. However, that condition rarely occurs.

IMPORTANT RULES FOR TAKING THE CET TEST

The rules that are given in this section apply to both the Associate-level and the Journeyman-level Tests. You should read these rules carefully. They are printed on the back of each test, but you do not want to waste valuable time reading instructions when you are ready to sit for the test.

■ Do not write in test book. Use the answer sheet to record your answers. You can use the back of the answer sheet for making calculations.

■ There are no "trick" questions in this exam. Do not look for questions with two or more right answers or questions in which there are no correct answers.

■ The purpose of the exam is to test your general knowledge of electronics and troubleshooting procedures. It does not require a specialized knowledge of the peculiarities of a particular brand-name system. Therefore, do not answer the questions according to some special case that you may know of.

■ If more than one answer appears to be correct, select the one that is most correct without any qualifications. For example, consider this question:

Raincoats are
(1) yellow.
(2) waterproof.
(3) plastic.
(4) obsolete.

To answer this question properly, choice (2) is *most* correct. It is true that raincoats may or may not be yellow, plastic, or obsolete. However, they are always supposed to be waterproof, so that it is the most correct answer of the choices.

■ You must correctly answer 75% (or more) of the questions to pass this exam. Educated guesses are appropriate. In other words, you are not penalized extra points for wrong answers. So, if you cannot decide between choices, pick the one you think is the best choice. Do not leave the answer blank because that will mean you have no chance to get credit for that question.

■ Read and answer each question carefully. Carelessness can cost you valuable grade points.

■ There is no official time limit. However, it would seem to be pointless to go beyond 4 hours. Typically, it will take about one hour to complete your answer sheet.

■ You may use a calculator provided it is *not* programmable or one with alphanumeric readouts.

■ No notes or books may be used.

■ No talking during the test.

When you are ready to take the test, write to ISCET. They will provide a list of certification administrators in your area. (Be sure to give your phone number when you write.) The address of ISCET is given in the Introduction of this book.

PROGRAMMED REVIEW

Start with Block number 1. Pick the answer that you feel is correct. If you select choice number 1, go to Block 13. If you select choice number 2, go to Block 15. Proceed as directed. There is only one correct answer for each question.

BLOCK 1

In order to successfully pass a CET test, you must answer
(1) at least 75% of the questions correctly. Go to Block 13.
(2) at least 70% of the questions correctly. Go to Block 15.

BLOCK 2

Your answer to the question in Block 8 is not correct. Go back and read the question again and select another answer.

BLOCK 3

The correct answer to the question in Block 13 is choice (1). Since the first 75 spaces on the answer sheet are for the Associate-level Test, the answer for the first question in the Journeyman-level Test will go in the space marked 76. Questions in the Journeyman-level Test are always numbered 76 through 150.

Here is your next question:

Answer sheets are

(1) always numbered vertically. Go to Block 6.
(2) always numbered horizontally. Go to Block 16.
(3) numbered either vertically or horizontally, and you should be careful before starting the test that you observe the type of numbering on the sheet. Go to Block 10.

BLOCK 4

Your answer to the question in Block 10 is not correct. Go back and read the question again and select another answer.

BLOCK 5

Your answer to the question in Block 7 is not correct. Go back and read the question again and select another answer.

BLOCK 6

Your answer to the question in Block 3 is not correct. Go back and read the question again and select another answer.

BLOCK 7

The correct answer to the question in Block 14 is choice (2).

Here is your next question:

Which of the following statements is correct?

(1) Only job-related experience applies to the journeyman requirement for experience. Go to Block 5.
(2) You may apply some of your formal schooling toward the experience required for a journeyman rating. Go to Block 17.

BLOCK 8

The correct answer to the question in Block 17 is choice (1).

You will need pencils, and that is all you will need (except for your highly specialized knowledge). (Using a pencil makes it possible to change an answer.) The papers are not machine graded, but you should make your marks clear and dark.

Here is your next question:

Which of the following statements is correct?

(1) There is only one correct answer to each question in the CET test. Go to Block 18.

(2) It is possible to have two or more correct answers for each question in the test. Go to Block 2.

BLOCK 9

Your answer to the question in Block 14 is not correct. Go back and read the question again and select another answer.

BLOCK 10

The correct answer to the question in Block 3 is choice (3). Don't get careless! Answer the questions carefully, and watch where you put your answers!

Here is your next question:

When you are getting ready to take the CET test

(1) you should tell the examiner (who is usually called the administrator) which type of equipment you work on so that the test questions will be related to the equipment you are familiar with. Go to Block 4.

(2) you will be given a test that does not require a specialized knowledge of any particular manufacturer's equipment. Go to Block 14.

BLOCK 11

Your answer to the question in Block 13 is not correct. Go back and read the question again and select another answer.

BLOCK 12

Your answer to the question in Block 17 is not correct. Go back and read the question again and select another answer.

BLOCK 13

The correct answer to the question in Block 1 is choice (1). To get a passing grade of 75% you cannot miss more than 18 questions out of a total of 75.

Here is your next question:

Your first answer for the Journeyman-level Test should be put in the space for

(1) question 76. Go to Block 3.

(2) question 1. Go to Block 11.

BLOCK 14

The correct answer to the question in Block 10 is choice (2). In some cases a manufacturer's drawing is used as part of a test. The questions asked about that drawing, however, do not require a special knowledge of the equipment.

Here is your next question:

To become a journeyman CET, you must have

(1) ten years experience in your field. Go to Block 9.

(2) four years experience in your field. Go to Block 17.

BLOCK 15

Your answer to the question in Block 1 is not correct. Go back and read the question again and select another answer.

BLOCK 16

Your answer to the question in Block 3 is not correct. Go back and read the question again and select another answer.

BLOCK 17

The correct answer to the question in Block 14 is choice (2). The staff at the ISCET office is always willing to correspond with you regarding any questions you may have about the CET test.

Here is your next question:

Which of the following statements is correct?

(1) No notes or books may be used when you are taking the CET test. Go to Block 8.

(2) You may use this book for reference when you are taking the CET test. Go to Block 12.

BLOCK 18

The correct answer to the question in Block 8 is choice (1).

Here is your next question:

Is the following statement correct?

If you can't decide between two statements, you should never guess because, if you're wrong, it will cost you extra points.

______________________. Go to Block 19.

(correct or not correct)

BLOCK 19

The statement in Block 18 is *not* correct. You are not penalized for wrong answers. If you can't decide between two of the choices, pick the one you think is right. Do not leave blank spaces on your answer sheet. Those blank spaces will surely be counted against you.

You have now completed the programmed section.

Two-Terminal Components

There is no section in the Journeyman Consumer Test called *Components and Circuits.* However, you must have a good understanding of these subjects in order to be able to answer questions in other sections of the test.

In the subject of electronics, a study of components is usually used as a foundation for the more complicated subjects. Following a discussion of components, the next subject to be taken up is usually circuits. They can be described as combinations of components. Finally, the circuits are combined into systems, which is the most complicated electronics subject. Systems can be defined as combinations of circuits.

A review of the first two subjects (components and circuits) is covered in this chapter. In the rest of the book you will find some material on systems.

If you are looking at the overall structure of the CET test, the Associate-level Test is primarily concerned with components and the simpler circuits, while the Journeyman-level Tests have questions about advanced circuits and complete systems. However, this is not a strict delineation because you will find some power supply systems discussed in the Associate-level Test, and you are sure to find questions on components and basic circuits in the Journeyman-level Test.

LINEAR, BILATERAL CIRCUIT ELEMENTS

A *linear* circuit component is one that follows Ohm's law. For example a resistor is an example of a typical linear component. Specific examples are: the carbon resistor, the metal film resistor, and the wire-wound resistor. There are other resistor components that do not follow Ohm's law, and therefore, are nonlinear. Examples are: the thermistor and the varistor.

A circuit component is *bilateral* if it will conduct equally well in either direction. A resistor is bilateral, whereas a diode is a unilateral component. Resistors, capacitors, and inductors are considered to be the most common linear bilateral circuit elements.

Resistors

The most common uses of resistors in electronics circuits are:

- Limit current
- Introduce a voltage drop
- Generate heat

Several years ago while I was on a lecture tour I made the statement that resistors were used only for those purposes. After one of the lectures, a technician from the audience came up and handed me a peaking coil and stated quite factually that "resistors are also used as coil forms." (A peaking coil is a resistor in parallel with an inductor. It is used primarily for the purpose of increasing the high-frequency response of broad-band amplifiers—specifically, the video or luminance amplifier in a television receiver.)

After conceding that point (reluctantly), another technician approached me and explained that one use of resistors that doesn't fall under those three categories is as noise generators. All resistors produce noise when electric current flows through them. It is known as thermal *agitation noise* or *Johnson noise*. The cause is the random motions of electrons in the resistor at room temperature. Thermal agitation noise is an example of white noise because it does not have any major frequency emphasis. I finally came to the conclusion that the best way to state the uses of resistors is to say that the three listed above are the most common uses.

When you go to purchase a resistor you must know the resistance value, and you must also know the power rating. Technically there is a third rating you should know if you are a designer, and that is the voltage rating of the resistor. When some resistors—such as the carbon composition type—are located in strong electric fields, they undergo a change in resistance. Also, they may be subject to voltage stress between the resistor and adjacent components.

The *carbon-composition resistor* is by far the most popular type. The reason is easy to understand: it is the cheapest type available today. At one time you could buy them in 5%, 10%, and 20% tolerances, but at the present time the 20% tolerance resistors are not being manufactured. Nevertheless, there are many of them still in use.

Carbon resistors can be obtained with power ratings from one-eighth watt to two watts, and they are usually made with a five-color code in the form of bands around the resistors. Naturally, if you are ready to take the Journeyman-level Test you know the resistor color code and you are aware of how to use it. Experienced technicians who take the Associate-level Test often miss the color codes for one-ohm and ten-ohm resistors. To a lesser extent the 100-ohm resistor also seems to cause them trouble.

The color code for resistors less than 10 ohms and greater than 1 ohm and also those less than 1 ohm are also important if you are planning to take the Associate-level Test. You may also encounter questions that require a knowledge of color codes when you take the Journeyman Consumer Test.

Typical CET Test Question

In the circuit of Fig. 2-1 resistor R is part of a power supply decoupling filter. You would expect this resistor to be color coded:

(1) Brown, black, green, silver
(2) Brown, black, brown, silver

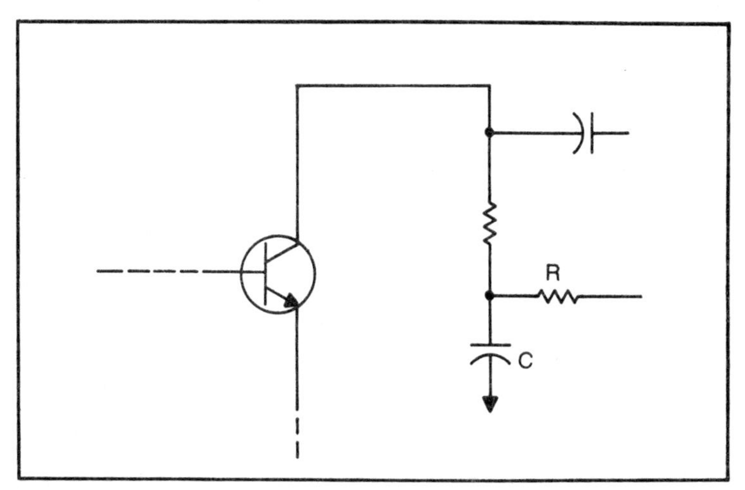

Fig. 2-1. What is the value of "R" in this circuit?

Answer: Choice 2 is correct. The resistance value represented by choice 1 is so large it would drop most of the supply voltage and starve the transistor amplifier.

In the five-color code the fifth band is used for reliability information. It tells the number of resistors that must be acceptable in a certain family of resistors. That information is primarily for designers and purchasing agents and has little to do with the technician. However, you should know the purpose of that band.

There are two distinct disadvantages of carbon-composition resistors. This first is that they produce a relatively high noise—especially when compared to some other resistance types available. This noise increases with an increase in temperature. Again, it is a white noise called thermal agitation or Johnson noise. To be technically accurate *the noise increases as the square root of the temperature increases.*

A second disadvantage of the carbon-composition resistor is that it has a relatively high temperature coefficient. This simply means that a small change in temperature can produce a relatively large change in resistance. This is especially true when the resistor is operated in the zero- to 60-degree Celsius range.

A positive temperature coefficient exists in that region. In other words, the resistance increases as the temperature increases.

Metal film resistors are more expensive than the carbon-composition type. They are available in resistance values of 0.1 ohm to 1.5 megohm. By comparison, the carbon-composition resistors have values in the range of 1 ohm to over 20 megohms.

Metal film resistors are normally available in one-tenth to one-watt range, and they are less popular than composition types simply because they are more expensive. An important feature of the metal film resistor is its low temperature coefficient. Another important feature is its low noise. Therefore, the two disadvantages of the carbon-composition resistor are not a major problem with metal film types.

Carbon film resistors are available in resistance values as high as 100 megohm, and tolerances from 0.5% up. On the lower end of the available resistance range, carbon film resistors produce a relatively low noise compared to the carbon-composition type. They have a negative temperature coefficient meaning that as the temperature increases their resistance value decreases. You will find them used most often in places where a very high resistance value is needed, or where the negative temperature coefficient characteristic is important.

Wire-wound resistors are more expensive than carbon-composition type and they are available off the shelf in one-half to three-watt power ratings. They are made by winding a resistance wire on a nonconducting form. Normally this would cause the component to act as both a resistance and also an inductance. A special bifilar (noninductive) winding can be used if the self inductance of the component is a serious consideration. An example of a noninductive winding is shown in Fig. 2-2.

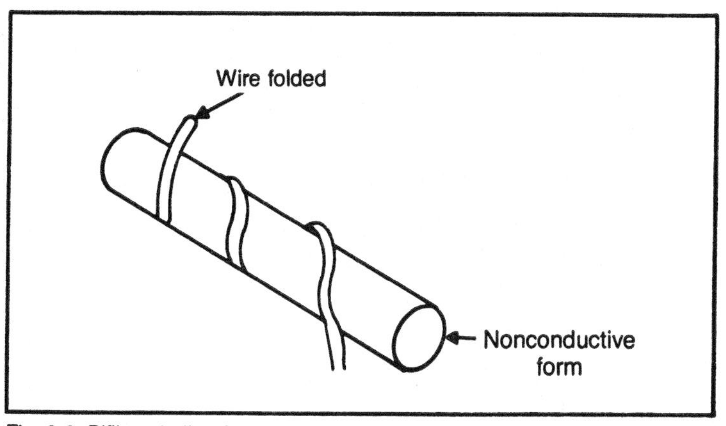

Fig. 2-2. Bifilar winding for wire-wound resistors.

Wire-wound resistors can be purchased in 1% to 10% tolerance range. Resistors can also be purchased in integrated circuit arrays. They look like any other DIP integrated circuit packages, but inside there are series- and parallel-connected resistors. The advantage of the integrated circuit configuration is a saving in space and also the saving in expense by using a single component for a number of different purposes.

As a journeyman technician you are expected to be able to calculate the resistance of series and parallel resistor networks. (The same is true for series and parallel capacitors. However, you will not encounter series and parallel inductors because of the problems introduced by mutual inductance.)

The most popular variable resistors are either carbon-composition types or wire wound. These components are connected into the circuit either as *potentiometers* or *rheostats*. The difference between these connections is that a potentiometer is a three-terminal network used to control the voltage of the circuit. A rheostat is a two-terminal connection used to control the circuit current. These two connections are shown in Fig. 2-3.

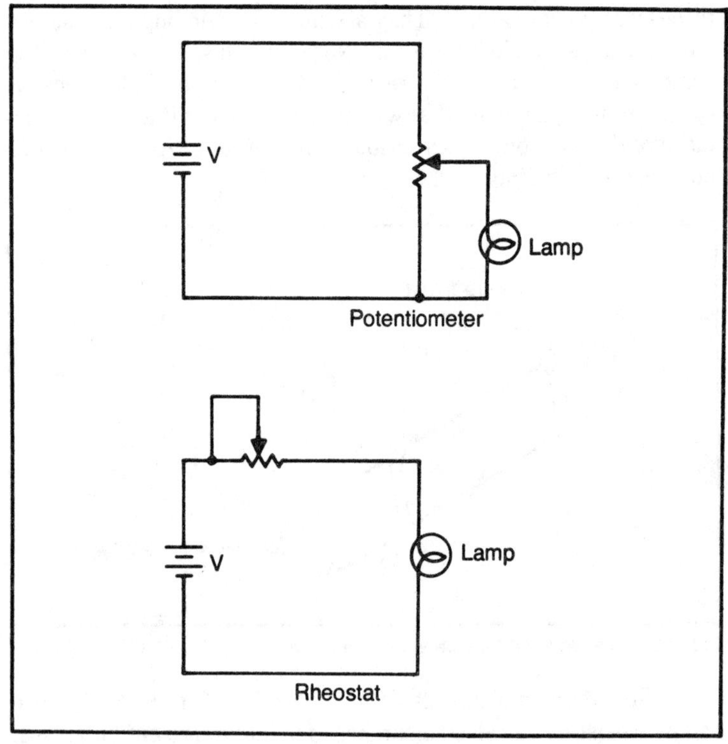

Fig. 2-3. Variable resistors are used in two different ways—electric and electronic circuits.

Note: The term potentiometer is not only used for a variable resistor but it is also used for an accurate voltage measuring device. Many technicians have missed questions about potentiometers (the voltage measuring type) in Journeyman-level CET Tests. These questions are normally found in the Industrial Journeyman Option.

Capacitors

Capacitors are used in electronic circuits for:

- Storing energy
- Introducing a voltage drop
- Producing a low-opposition path to high frequencies

Many technicians have argued with me that there is really only one use of a capacitor and that is to store energy. Their argument is very convincing. I list three purposes because they make it easier to understand applications in various circuitry.

At one time *paper capacitors* were very popular. These were capacitors that used a wax paper dielectric. That type is no longer used extensively for several reasons. One is their very wide temperature coefficient and the other is the difficulty in getting a good insulation between the plates.

The capacity (which may be called capacitance in the test) of a capacitor is determined by the distance between the plates, the area of the plates, and most important, the type of dielectric.

Instead of wax paper, material with much higher dielectric constants and much better insulating qualities are now used for inexpensive capacitors. One example is the *Mylar film* type. Other excellent dielectric materials are polystyrene, polycarbonate, polypropylene, and Teflon. The advantage of the Teflon is its very high-voltage breakdown rating.

Ceramic capacitors are very popular in modern electronic circuits. Temperature compensating ceramic capacitors can be purchased with positive or negative temperature coefficients. Also, they are available with a zero temperature coefficient. The temperature coefficient rating is identified by a number such as N220 (which is a capacitor with a negative temperature coefficient) and P750 (which is a capacitor with a positive temperature coefficient). The number tells the number of parts per million change in capacity with a one degree Celsius change in temperature.

If a capacitor is rated NPO, it has no appreciable change in capacity for a given change in temperature. When you are replacing a ceramic type capacitor it is very important that you get the right temperature coefficient. If you put a positive temperature coefficient in place of a negative type, it can cause serious problems. For example, in an oscillator circuit it can cause the oscillator to drift over a wide range of frequencies for relatively small changes in temperature.

Mica capacitors are used in places where a high dielectric breakdown voltage is important. These capacitors are especially useful at high frequencies. They are made in two major categories. The aluminum type is the older and lower-cost type. The tantalum type is more expensive, but has a longer shelf life. Remember, however, that tantalum is *highly poisonous* and some environmentalists question the need for tantalum capacitors in circuits. The biggest concern is the method of discarding these capacitors. *You should never discard them in such a way that they can ultimately find their way into the soil or water table.*

An interesting feature of electrolytic capacitors is their wide range of capacitance values in the upper range of capacity. It is not uncommon to have electrolytic capacitor values anywhere from a range of one microfarad to 100,000 farads.

Nonpolarized electrolytic capacitors are available. They can be charged, but they normally have less capacity than the polarized types.

Variable capacitors are usually considered to have an air dielectric. In the smaller types it is common practice to use some kind of a thin plastic film between the plates to prevent them from shorting out, but they are still considered to be air dielectric types.

One time I was lecturing on capacitors and I explained to my students that *the energy of a capacitor is always stored in the dielectric.* One of the students asked this rather challenging question: "If the energy is stored in the dielectric, does that mean that it would be possible to change stations in a radio by blowing out the air between the plates of the tuning capacitor?" The answer, of course, is that you cannot blow out the air between the plates of a capacitor. However, the question was very interesting.

In terms of variable capacitors you should remember that *padders* are used in series with variable capacitors to change the capacitance range. Trimmers, on the other hand, are used in parallel with a variable capacitor to change the range.

Inductors

The most important application of inductors in electronics is in tuned circuits and filter circuit configurations. They are used to resonate with capacitors. In some rare circuits, inductors may be used in combinations with resistors in order to produce time constant circuits. Remember that the time constant of an RC circuit is $T = RC$, whereas, the time constant of an inductor circuit is $T = \frac{L}{R}$.

Transformers, which are inductive components, are used for matching impedances. Keep in mind the fact that special transformers such as the balun (discussed in the chapter on antennas and transmission lines) are used extensively in consumer electronic products. Gyrators are ICs used to simulate inductance in circuits.

Typical CET Test Question

The dots on the transformer symbol in Fig. 2-4 mean

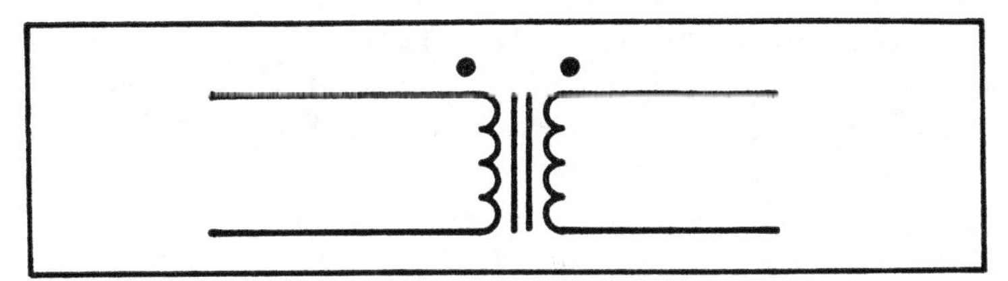
Fig. 2-4. The dot notation marks the in-phase leads on the transformers.

(1) the windings must be grounded at those points.
(2) the windings must *not* be grounded at those points.
(3) (Neither choice is correct.)

Answer: Choice 3 is correct. The dots show the points that are in phase.

NONLINEAR TWO-TERMINAL COMPONENTS

The most common of the nonlinear two-terminal components are the diodes. This includes both the vacuum tube and semiconductor types. In order to stay aligned with the modern technology, the number of questions on vacuum-tube diodes is greatly reduced. However, there are some vacuum-tube diodes which must be taken into consideration in the overall picture. For example, a *magnetron* is a form of a diode. (Magnetrons are used in microwave heating devices.)

The neon lamp is technically not a vacuum-tube device. Instead, it is a gas-filled device. Nevertheless, you might find a question regarding neon lamps in the CET test. Remember that the neon lamp has been virtually replaced by the *diac*, which is a semiconductor two-terminal breakover device.

The characteristic curves of these two devices (neons and diacs) are very similar. They breakover in both directions, and they will not conduct until the breakover point is reached. In the neon lamp the breakover point is sometimes referred to as the *firing potential.* Once they breakover, they have a relatively constant voltage despite changes in current through them. That makes them useful—especially the neon lamp—as simple voltage regulators.

Semiconductor diodes may be connected in series or parallel as shown in Fig. 2-5. When they are connected in series as shown in Fig. 2-5(A) their peak inverse voltages (PIV) add. So, they are series-connected in applications where the peak inverse voltage rating of one diode is not sufficiently high for the reverse voltage encountered in the circuit.

Resistors are connected across the series-connected diodes in order to equalize the reverse voltage. This is necessary because the reverse resistance of the diodes are not exactly the same values. Capacitors may also be connected across a diode to reduce the possibility of damage from transient voltages.

Diodes are connected in parallel as shown in Fig. 2-5(B) in order to increase their current-delivering capacity. Starting resistors are connected in series with the parallel-connected diodes. This is to assure that each diode will conduct and carry its share of the current.

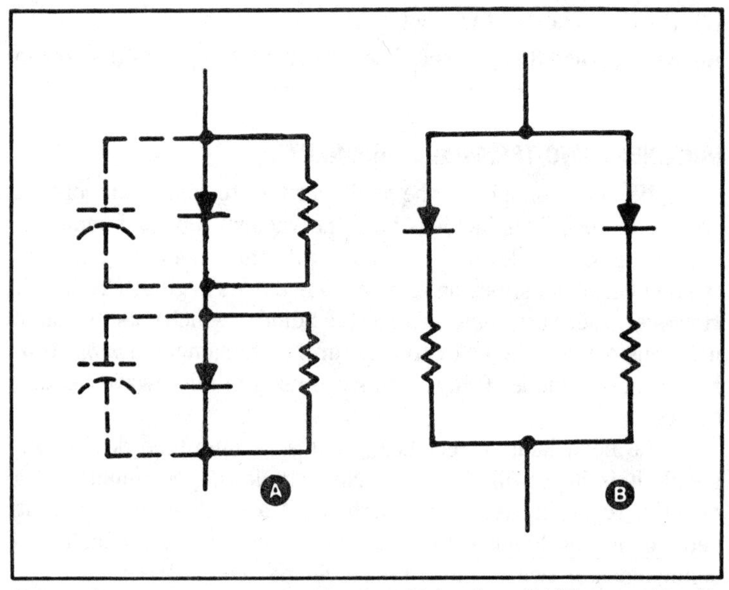

Fig. 2-5. Diodes may be combined to obtain high ratings. (A) When combined in series their PIV ratings are additive. (B) When connected in parallel their current ratings are additive.

Remember that in semiconductor diode packages it is the cathode section that is always marked. This mark may be a band around one end of the diode package, or it may even be a positive sign. Regardless of the method used, the marking is always on the cathode side.

Constant-current diodes, as their name implies, provide a constant current flow under varying voltage conditions. One application of constant current diodes is to linearize the sawtooth waveform in a typical time constant. An example is shown in the UJT circuit of Fig. 2-6. During the charge time of the capacitor, which produces the long ramp on the sawtooth waveform, the

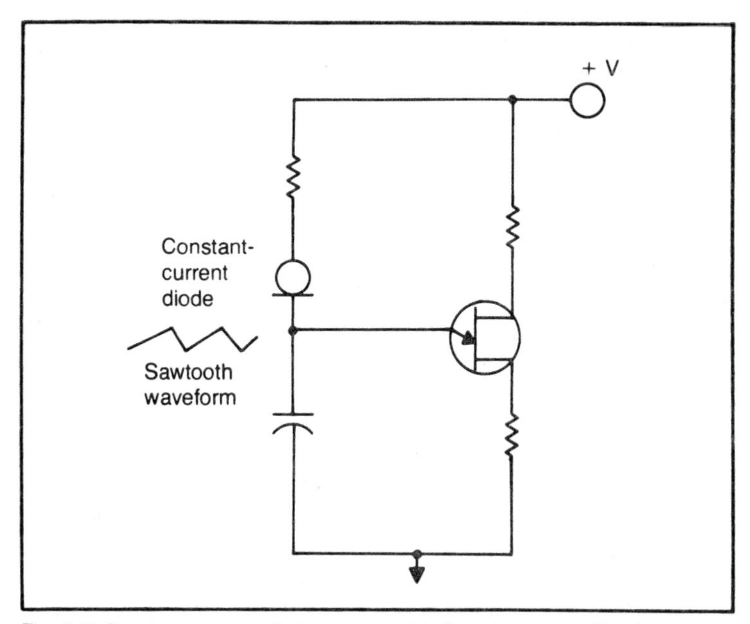

Fig. 2-6. Constant-current diodes are used to linearize sawtooth voltages.

charging current is linearized by the constant-current diode in the circuit.

Constant-current diodes are usually made by taking a JFET and tying the gate to the drain. That makes it a two-terminal device.

You have to be careful with the symbol for the constant-current diode. It looks very much like the symbol for the tunnel diode shown in Fig. 2-7. However, the two devices are nothing alike. There are several other symbols for tunnel diodes shown in the same illustration.

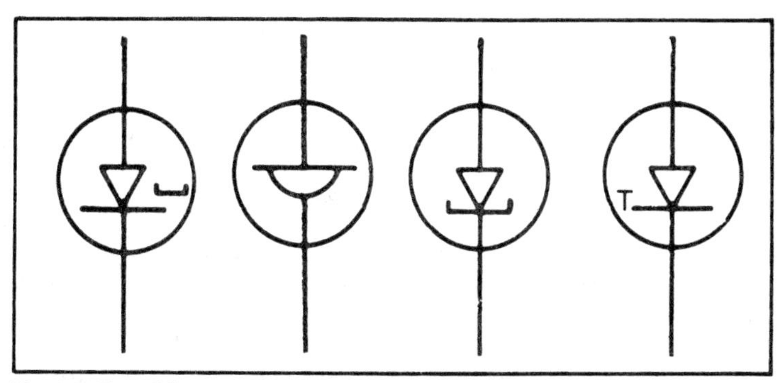

Fig. 2-7. Four different schematic symbols are used to represent tunnel diodes. Notice how much the third one looks like a constant-current diode.

A tunnel diode is the fastest-acting switch in electronics, and that is one of its most important applications. It is also used as an ultra-high frequency oscillator.

Two examples of *optoelectronic diodes* are the LED (Light-Emitting Diode) and the LAD (Light-Activated Diode). When the LED is conducting and producing light there is a voltage across it of aproximately 1.6 volts. In practice this may vary from 1.5 to 2 volts but it is usually within this range. Manufacturers usually give the voltage drop as 1.6 to 1.7 volts. You can get an LED to light with other voltages, but the manufacturer's specification is based on having the LED give off a specified amount of light. It is also based on maintaining the operation of the LED over a long period of time.

Light-activated diodes are open switches in the absence of an input light. When light strikes them, they conduct in the forward direction, but not in the reverse direction.

A device that may use both an LED and an LAD is shown in Fig. 2-8. It is called an *optical coupler*. Its advantage is a very high isolation resistance between the input and output circuits. In place of the LAD, some optical couplers use phototransistors or other optoelectronic devices. In any event, keep in mind the fact the isolation resistance between the input and output circuits is extremely high, making it possible to interface circuits that operate at two different voltage or current levels.

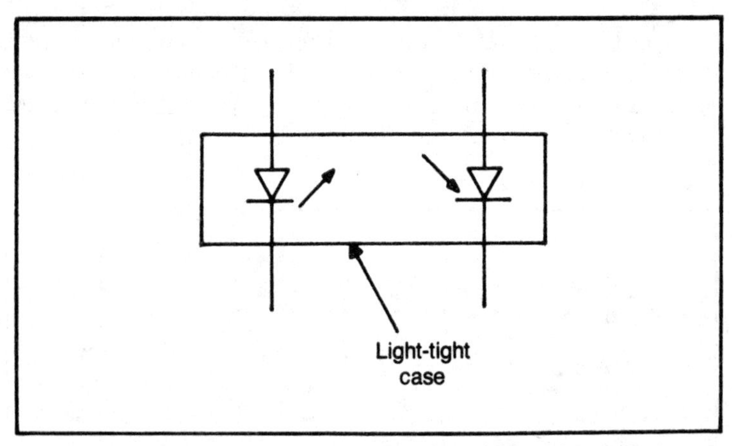

Fig. 2-8. An optical coupler can be made with an LED and an LAD that are light-coupled.

Optical couplers have also been used as variable resistors in volume control and tone control circuitry. They have the advantage that they produce very little noise during changes between the input and output levels.

The *four-layer diode* shown in Fig. 2-9 is also known as a *Shockley diode*. Its characteristic curve shows that it is a breakover device in one direction, but a simple diode in the opposite direction. Do not confuse it with a *Schottky diode,* which is made by interfacing semiconductor and metal materials at a junction. They are also called *hot-carrier diodes*. These diodes are characterized by a high conducting current capability, and very low forward voltage drop.

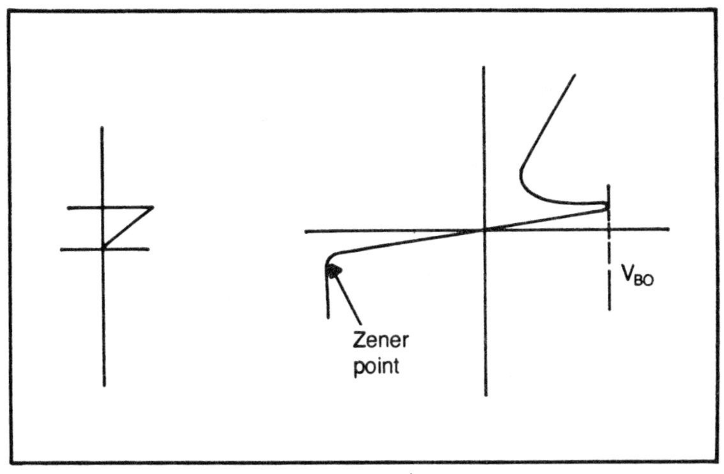

Fig. 2-9. The schematic symbol for a four-layer diode and its characteristic curve. V_{BO} is the forward breakover point.

Varactor diodes behave like capacitors in circuits. A reverse voltage across the diode sets the thickness of the depletion region that acts like the capacitor dielectric. Remember that increasing the reverse voltage also increases the thickness of the dielectric. That, in turn, reduces the capacity.

Zener diodes, like varactors, operate with the reverse voltage. However, unlike the varactor, the zener is operated with a reverse current. The most common application of zener diodes is in voltage regulator circuits.

Thermistors and varistors (also called VDRs) are examples of nonlinear two-terminal devices that are not diodes. Thermistors are temperature-sensitive resistors that are used for sensing heat. VDRs are voltage-dependent resistors. They have a high resistance when there is a low voltage across them; and, they have a low resistance when subjected to a high voltage.

Typical CET Test Question

Why is the VDR connected across the relay coil in the circuit of Fig. 2-10?

(1) To reduce the time required for relay contact closure
(2) To protect the transistor

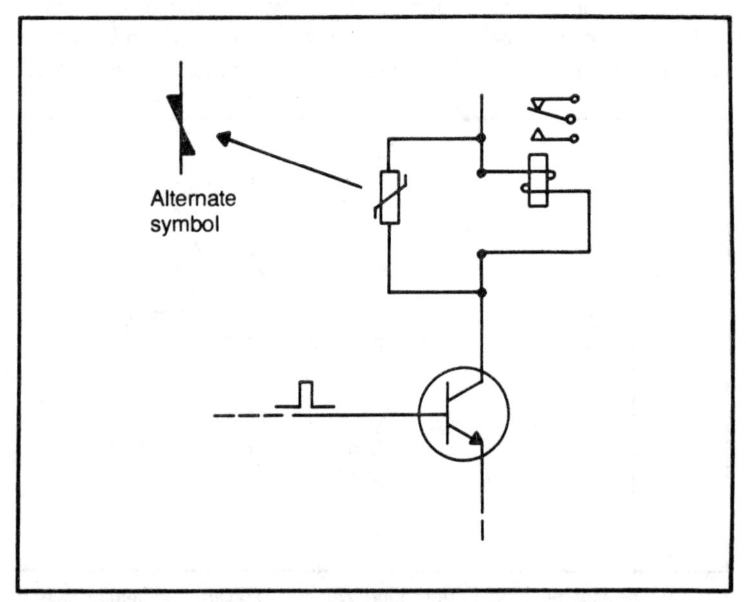

Fig. 2-10. What is the purpose of the component connected across the relay coil?

Answer: Choice 2 is correct. The input pulse momentarily energizes the relay coil. The inductive kickback resulting from pulsing the coil could destroy the transistor.

PROGRAMMED REVIEW

Start with Block number 1. Pick the answer that you feel is correct. If you select choice number 1, go to Block 13. If you select choice number 2, go to Block 15. Proceed as directed. There is only one correct answer for each question.

BLOCK 1

A buzzing power or audio transformer is usually the result of

(1) loose laminations. Go to Block 13.
(2) saturation. Go to Block 15.

BLOCK 2

The correct answer to the question in Block 32 is choice (2). Voltage is unit of *work* (*not* force)! When you move the plates apart you increase the distance necessary to move a unit charge between the plates. Since work is force times distance, increasing the distance will increase the work and therefore, it will increase the voltage.

Here is your next question:

Diodes are connected in series to get a greater

(1) PIV rating. Go to Block 8.
(2) current-carrying ability. Go to Block 26.

BLOCK 3

The correct answer to the question in Block 7 is choice (2). (Note: partition noise will be discussed in the next chapter.)

Here is your next question:

A capacitor (C) is connected in series with a tuning capacitor to change the range of capacity values. Capacitor C is called a

(1) trimmer. Go to Block 14.
(2) padder. Go to Block 22.

BLOCK 4

Your answer to the question in Block 21 is not correct. Go back and read the question again and select another answer.

BLOCK 5

Your answer to the question in Block 16 is not correct. Go back and read the question again and select another answer.

BLOCK 6

Your answer to the question in Block 12 is not correct. Go back and read the question again and select another answer.

BLOCK 7

The correct answer to the question in Block 12 is choice (1). As a general rule you can go to a component with tighter specifications, but you should never go the other way.

Here is your next question:

Another name for Johnson noise is

(1) partition noise. Go to Block 37.
(2) thermal agitation noise. Go to Block 3.

BLOCK 8

The correct answer to the question in Block 2 is choice (1). You are most likely to see series-connected diodes in high-voltage supplies.

Here is your next question:

The anode of the diode shown in this block is at

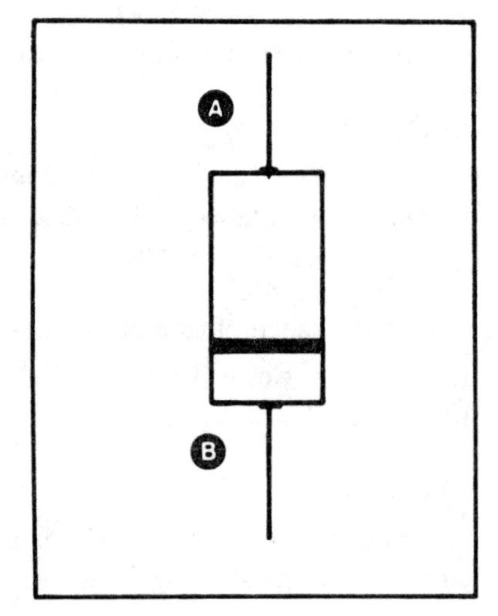

(1) side A. Go to Block 19.
(2) side B. Go to Block 35.

BLOCK 9

Your answer to the question in Block 16 is not correct. Go back and read the question again and select another answer.

BLOCK 10

Your answer to the question in Block 13 is not correct. Go back and read the question again and select another answer.

BLOCK 11

Your answer to the question in Block 25 is not correct. Go back and read the question again and select another answer.

BLOCK 12

The correct answer to the question in Block 30 is choice (2).

Here is your next question:

Is it OK to replace a carbon resistor with a metal film resistor? (Assume the resistance value and power rating are the same for both resistors. Disregard cost as a factor in making the decision.)

(1) It is OK to substitute a metal film resistor. Go to Block 7.
(2) This type of substitution should never be made. Go to Block 6.

BLOCK 13

The correct answer to the question in Block 1 is choice (1). Sometimes it is possible to tighten the laminations by tightening screws used to construct the transformer. If that isn't possible, the transformer must be replaced.

Here is your next question:

You would expect to see the component that has the symbol shown in this block used:

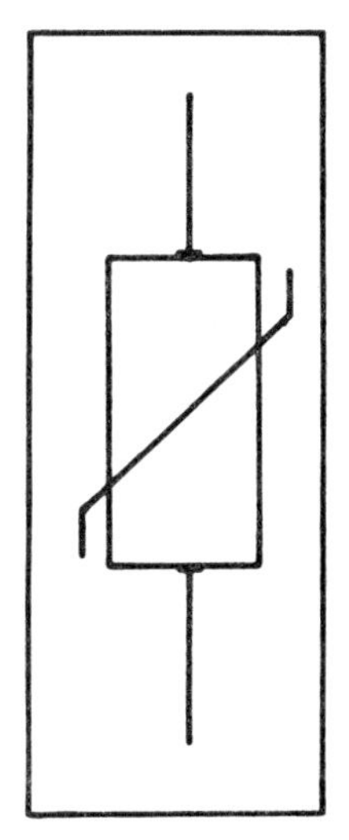

(1) As a parametric amplifier. Go to Block 10.
(2) To protect a transistor from an inductive kickback. Go to Block 20.

BLOCK 14

Your answer to the question in Block 3 is not correct. Go back and read the question again and select another answer.

BLOCK 15

Your answer to the question in Block 1 is not correct. Go back and read the question again and select another answer.

BLOCK 16

The correct answer to the question in Block 21 is choice (2).

There is no such rating as permanent temperature coefficient.

Here is your next question:

A capacitor with a temperature coefficient of N220 is used in a tunnel diode oscillator circuit. Which of the following statements is correct?

(1) It can be replaced with a capacitor having a rating of P220. Go to Block 9.
(2) It can be replaced with a capacitor having a rating of NPO. Go to Block 5.
(3) Neither choice is correct. Go to Block 30.

BLOCK 17

Your answer to the question in Block 36 is not correct. Go back and read the question again and select another answer.

BLOCK 18

Your answer to the question in BLOCK 30 is not correct. Go back and read the question again and select another answer.

BLOCK 19

The correct answer to the question in Block 8 is choice (1). The cathode side is always the one marked.

Here is your next question:

Which of the following operates with a reverse current?

(1) A varactor diode. Go to Block 29.
(2) A zener diode. Go to Block 24.
(3) Both choices are correct. Go to Block 33.

BLOCK 20

The correct answer to the question in Block 13 is choice (2). The symbol is for a varistor (*V*oltage v*AR*iable res*ISTOR*). It has a high resistance when the voltage across it is low, and a low resistance when the voltage across it is high. You will see this component connected across transistor-driven inductors and transformers, and in automatic degaussing circuits.

Here is your next question:

A bifilar winding might be used:

(1) For making a filter choke. Go to Block 31.
(2) For making a wire-wound resistor. Go to Block 36.

BLOCK 21

The correct answer to the question in Block 24 is choice (1). Decreasing the reverse bias has the effect of moving the plates closer together.

Here is your next question:

A capacitor with a temperature coefficient marking of P750 has:

(1) A *Permanent* temperature coefficient. Go to Block 4.
(2) A *Positive* temperature coefficient. Go to Block 16.

BLOCK 22

The correct answer to the question in Block 3 is choice (2).

Here is your next question:

Which of the following is an example of white noise?

(1) Thermal agitation. Go to Block 38.
(2) $\frac{1}{f}$ noise. Go to Block 34.

BLOCK 23

Your answer to the question in Block 32 is not correct. Go back and read the question again and select another answer.

BLOCK 24

The correct answer to the question in Block 19 is choice (2). Zener and varactor diodes both operate with a reverse voltage, but only the zener diode operates with a reverse current.

Here is your next question:

To increase the capacity of a voltage-variable capacitor

(1) decrease the reverse voltage. Go to Block 21.
(2) increase the reverse bias voltage. Go to Block 27.

BLOCK 25

The correct answer to the question in Block 36 is choice (1). Be sure you know all of the component symbols before taking any CET test.

Here is your next question:

Squeezing the turns of a coil closer together will

(1) decrease its inductance. Go to Block 11.
(2) increase its inductance. Go to Block 32.
(3) not affect its inductance. Go to Block 28.

BLOCK 26

Your answer to the question in Block 2 is not correct. Go back and read the question again and select another answer.

BLOCK 27

Your answer to the question in Block 24 is not correct. Go back and read the question again and select another answer.

BLOCK 28

Your answer to the question in Block 25 is not correct. Go back and read the question again and select another answer.

BLOCK 29

Your answer to the question in Block 19 is not correct. Go back and read the question again and select another answer.

BLOCK 30

The correct answer to the question in Block 16 is choice (3). Always replace ceramic capacitors with one having the same ratings.

Here is your next question:

The voltage drop across a glowing LED is usually about:

(1) 0.7 volts. Go to Block 18.
(2) 1.6 volts. Go to Block 12.

BLOCK 31

Your answer to the question in Block 20 is not correct. Go back and read the question again and select another answer.

BLOCK 32

The correct answer to the question in Block 25 is choice (2). When the windings are closer together there is a better electromagnetic coupling between them.

Here is your next question:

The capacitor shown in the illustration for this block has been charged to a voltage of 100 volts. Moving the plates of the capacitor further apart will

(1) decrease the voltage across the capacitor. Go to Block 23.
(2) increase the voltage across the capacitor. Go to Block 2.

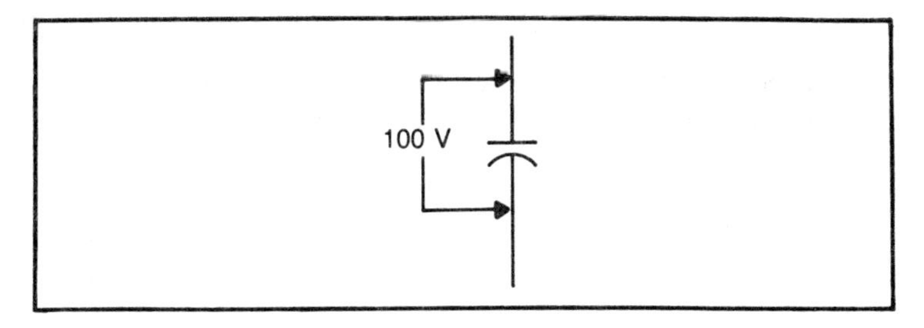

BLOCK 33

Your answer to the question in Block 19 is not correct. Go back and read the question again and select another answer.

BLOCK 34

Your answer to the question in Block 22 is not correct. Go back and read the question again and select another answer.

BLOCK 35

Your answer to the question in Block 8 is not correct. Go back and read the question again and select another answer.

BLOCK 36

The correct answer to the question in Block 20 is choice (2). A bifilar winding is noninductive, so it would be useless for making an inductor. It is used for making noninductive wire-wound resistors.

Here is your next question:

Which of the symbols in this block represents a constant-current diode?

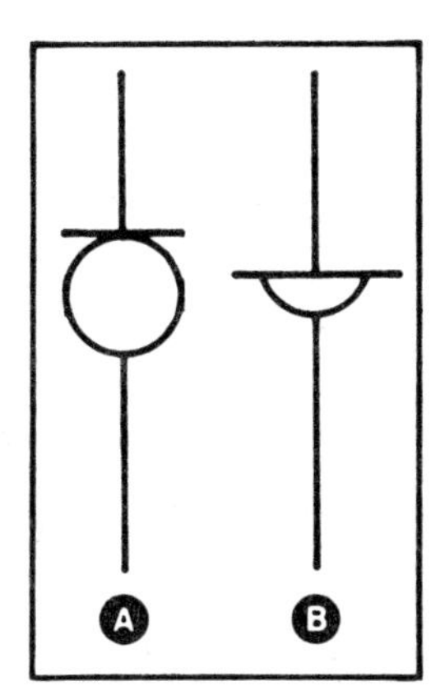

(1) The one shown in A. Go to Block 25.
(2) The one shown in B. Go to Block 17.

BLOCK 37

Your answer to the question in Block 7 is not correct. Go back and read the question again and select another answer.

BLOCK 38

The correct answer to the question in Block 22 is choice (2). White noise is evenly distributed over a wide range of frequencies.

Here is your next question:

What is the resistance between "A" and "B" in the integrated circuit array shown in this block?

______________ ohms.

Go to Block 39.

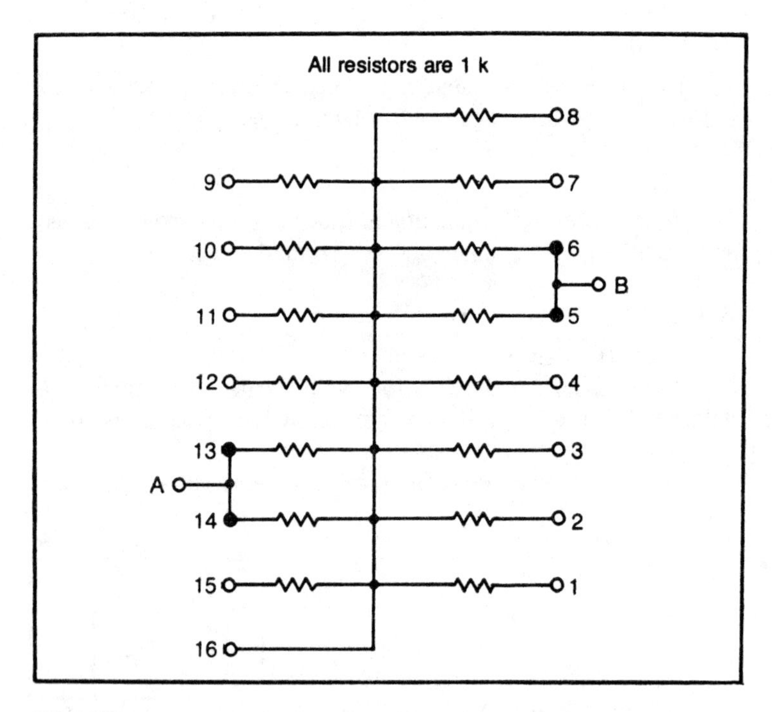

BLOCK 39

The resistance is 1 k.

You have now completed the programmed section.

ADDITIONAL PRACTICE CET TEST QUESTIONS

01. For a thermistor there is a wide variation in resistance corresponding to a relatively small change in

(1) time.
(2) temperature.
(3) humidity.
(4) voltage.

02. In order to be able to calculate the amount of current through a resistor by Ohm's law, it is necessary that the resistor be

(1) a VDR.
(2) a thermistor type.
(3) linear.
(4) unilateral.

03. Which of the following is *not* a linear resistor?

(1) Film
(2) Carbon composition
(3) VDR
(4) Wire wound

04. A certain resistor has a resistance value that depends upon the amount of light present. It is called

(1) a varactor.
(2) a PER.
(3) a PIR.
(4) an LDR.

05. In order to show that a resistor has a tolerance of ±10 percent

(1) the third band is silver.
(2) the fourth band is gold.
(3) the fourth band is silver.
(4) there is no color in the fourth band.

06. Select the proper color code for a 100-ohm resistor.

(1) Brown, black, brown
(2) Brown, black, black
(3) Brown, black, gold
(4) Brown, black, silver

07. Which of the following is not a common taper for variable resistors?

(1) Linear
(2) Audio
(3) Converse
(4) Reverse

08. A resistor with a linear taper
(1) has a code letter L on the case.
(2) is always wire wound.
(3) can be identified with an ohmmeter measurement.
(4) has a red dot near the center terminal.

09. To stabilize against temperature changes a circuit may use
(1) a thermistor.
(2) an LDR.
(3) a ferrite bead.
(4) a varistor.

10. Identify the polarized capacitor in the following list.
(1) Metallized paper
(2) Glass
(3) Ceramic
(4) Electrolytic

11. A certain capacitor has a rating of P150. This means that
(1) the capacitance will decrease 150 parts per million when the temperature increases 1° Celsius.
(2) the value of capacitance is 150 microfarads.
(3) the value of capacitance is 150 pF.
(4) none of these choices is correct.

12. Which of the following is an advantage of mica capacitors?
(1) Available in both low and high capacitance values.
(2) High breakdown voltage.

13. Variable capacitors may be purposely made nonlinear in order to get a linear dial on the radio. Is this statement true or false?
(1) True
(2) False

14. Which of the following is not an advantage of tantalum over aluminum oxide electrolytics?

(1) Longer shelf life
(2) Much lower cost

15. A measure of the opposition offered by a capacitor to the flow of dc is its

(1) ESR.
(2) Vdcw.
(3) Q.
(4) specific dielectric capacitance.

16. The reciprocal of dissipation factor capacitor is

(1) ESR.
(2) Vdcw.
(3) specific dielectric capacitance.
(4) Q.

17. When two capacitors are placed in parallel, the breakdown-voltage rating of the combination is equal to

(1) the smaller of the two breakdown-voltage ratings.
(2) the total of the breakdown-voltage ratings of each capacitor.
(3) a value that is proportional to the reciprocal of the sums of the breakdown-voltage ratings of each capacitor.
(4) the average of the two breakdown-voltage ratings.

18. Three parallel resistors are each dissipating 3 watts of power. What is the power dissipation of the circuit?

(1) 1 watt
(2) 3 watts
(3) 9 watts
(4) Cannot be determined from the information given.

19. Two 5-microfarad capacitors are connected in series. Their combined capacity is

(1) 2.5 microfarads.
(2) 3.3 microfarads.
(3) 5 microfarads.
(4) 10 microfarads.

20. A certain resistor is color-coded orange, orange, orange, gold. By actual measurement its resistance is 34.8 k.

(1) The resistor is out of tolerance.
(2) The resistor is in tolerance.

21. Two coils have equal lengths, equal radii, and the same number of turns. They are made with slightly different wire sizes. Which of the following is correct?

(1) They will both have the same inductance value.
(2) The one with the larger wire size will have the larger inductance value.
(3) The one with the smaller wire size will have the larger inductance value.

22. The inductance of a coil does *not* depend upon

(1) the number of turns in the coil.
(2) the current flowing through the coil.
(3) the distance between the turns of the coil.
(4) the shape of the coil.

23. The capacity of a 4-microfarad capacitor in series with a 6-microfarad capacitor is

(1) 2.4 microfarads.
(2) 3.1 microfarads.
(3) 10 microfarads.
(4) 24 microfarads.

24. The capacity of a capacitor is not affected by

(1) the type of material used for the dielectric.
(2) the area of the plates facing.
(3) the type of material used for the plates.
(4) the distance between the plates.

25. If the collector current of a transistor flows all of the time, the transistor is being operated

(1) class-A.
(2) class-B.
(3) class-C.

26. The efficiency of a power supply that is delivering maximum power is

(1) minimum.
(2) maximum.

(3) 50%.
(4) 0%.

27. The maximum power that a battery can deliver to a load resistor occurs when

(1) there is a short circuit across its terminals—that is, when $R_L = 0$.
(2) there is an open circuit across its terminals—that is, when $R_L = \infty$.
(3) the load resistance is adjusted to equal the internal resistance of the battery.
(4) cannot determine the answer from the information given.

28. Electron current flows in a PNP transistor

(1) from emitter to collector.
(2) from collector to emitter.

29. Which of the following types of bias is not used with transistors?

(1) AGC bias
(2) Power-supply bias
(3) Battery bias
(4) Self bias using an emitter resistor

30. The base voltage of an NPN transistor is

(1) positive with respect to the collector voltage.
(2) negative with respect to the collector voltage.

31. Which is the most efficient class of amplifier?

(1) Class-A
(2) Class-AB2
(3) Class-B
(4) Class-C

32. Transistors are seldom operated

(1) class-A.
(2) class-AB.
(3) class-B.
(4) class-C.

33. Doubling the dc voltage across a thermistor will cause the current to

(1) double.
(2) be halved.
(3) change, but it is not possible to determine the current value by Ohm's law since a thermistor is a nonlinear component.

34. A certain resistor changes its resistance value in accordance with the amount of voltage across its terminals. It is called

(1) a VDR.
(2) a VER.
(3) a VIR.
(4) an EDR.

ANSWERS TO PRACTICE TEST

Question Number	Answer Number
01	(2)
02	(3)
03	(3)
04	(4)
05	(3)
06	(1)
07	(3)
08	(3)
09	(1)
10	(4)
11	(4)
12	(2)
13	(1)
14	(2)
15	(1)
16	(4)
17	(1)
18	(3)
19	(1)
20	(1)
21	(1)
22	(2)
23	(1)
24	(3)
25	(1)
26	(3)
27	(3)
28	(2)
29	(4)
30	(2)
31	(4)
32	(4)
33	(3)
34	(1)

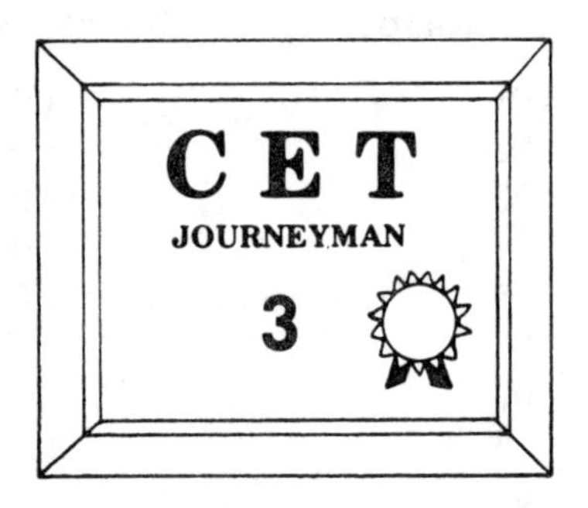

Three-Terminal Amplifying Components and Basic Circuits

In Chapter 2 you reviewed some of the characteristics of nonelectronic components specifically resistors, capacitors, and inductors. You also reviewed some of the electronic two-terminal devices.

In this chapter some three-terminal amplifying and switching components will be reviewed. Some basic electric circuits for two-terminal linear components will also be included.

BASIC ELECTRIC CIRCUITS

There are some combinations of two-terminal linear bilateral circuits elements that you should review before taking the Journeyman-level CET Test. Figure 3-1 shows four basic filter configurations that use these components. If you keep in mind the fact that inductors tend to pass low frequencies and reject high frequencies, while capacitors tend to pass high frequencies and reject low frequencies, it is an easy matter to identify which type of filter you are looking at. For example, in the low-pass filter the inductor—which is in series between the generating end (left side) and the load end (right side)—will pass the low frequencies from the generator to load. The capacitors which are across the line bypass the high frequencies to ground around the load.

The characteristics of series- and parallel-tuned circuits are also important. They are summarized in Fig. 3-2. Be especially careful to note that in the parallel-tuned circuit *the resonant frequency can be selected by adjusting the variable resistor in either of the*

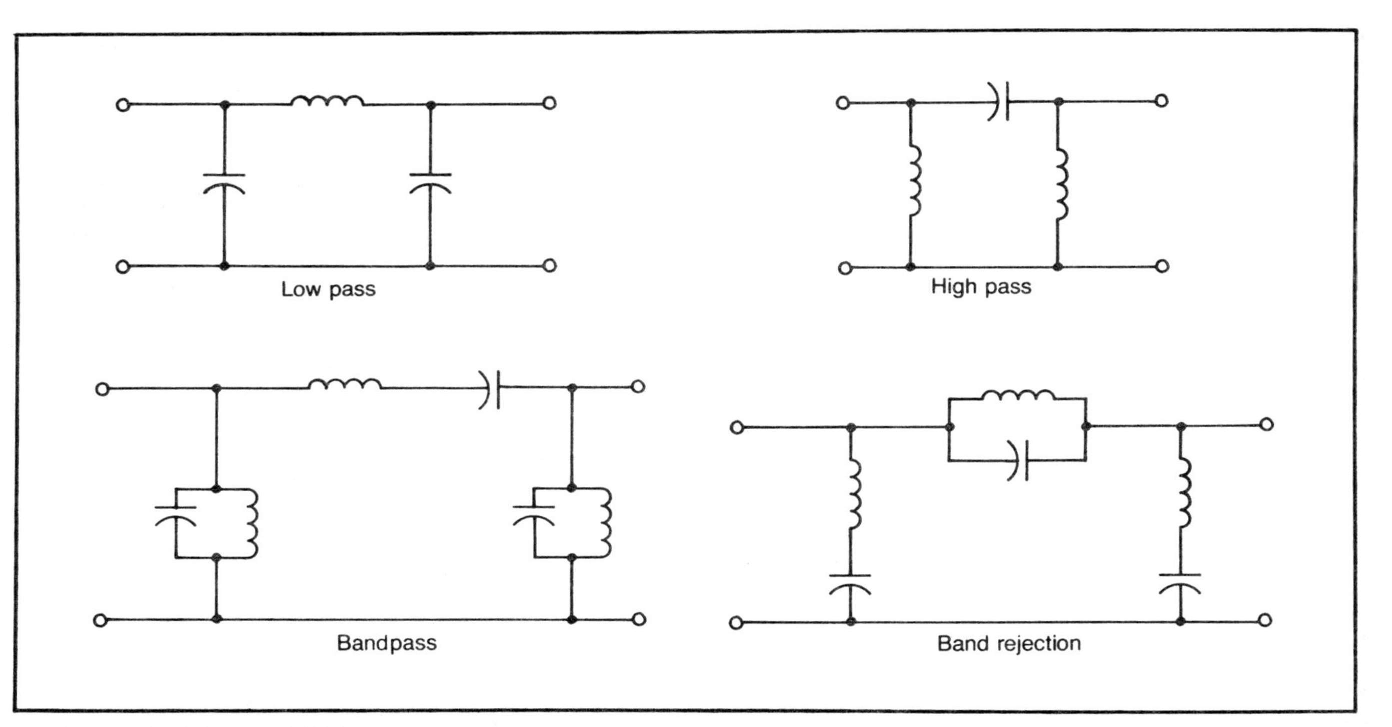

Fig. 3-1. Four basic types of LC filters.

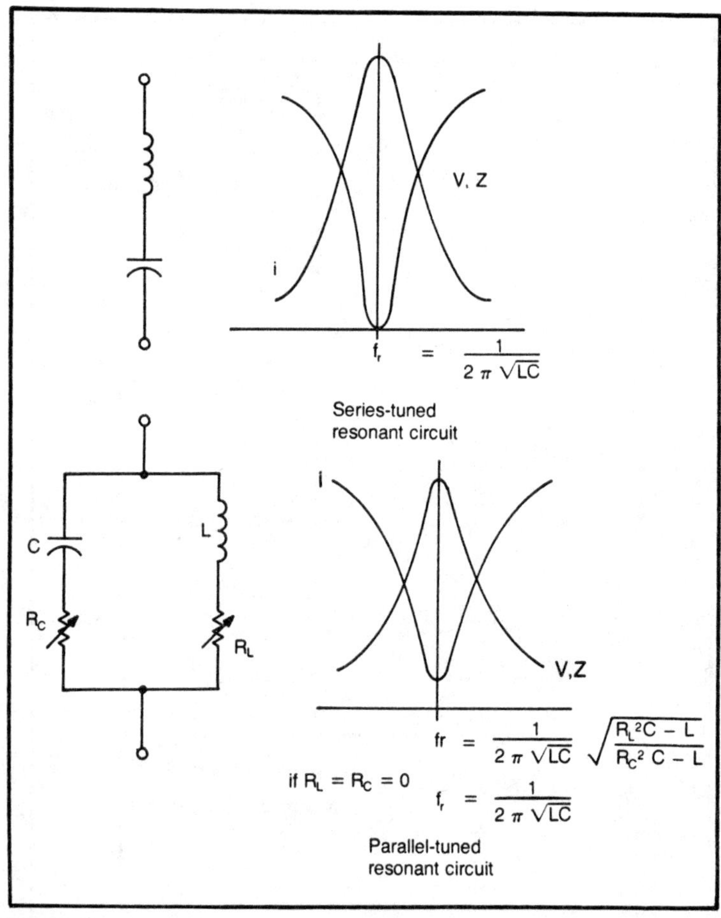

Fig. 3-2. Characteristics of series and parallel-tuned circuit. L or C can be used to select the resonant frequency. In the parallel-tuned circuit the resonant frequency may also be selected by R_L or R_C.

branches! In this sense, parallel-tuned circuits are different from the series-tuned circuits that cannot be frequency adjusted by using a resistor.

The equations for resonant frequency and the characteristic curves for these tuned circuits are shown in the illustration. Note that in the parallel resonant equation if R_L and R_C can be considered to be zero ohms, then the equation is the same as for series-tuned circuits. This special condition is the one that is very often taken up in basic theory books. So, the concept of resistance tuning is unfamiliar to many technicians. However, traps have been designed that can be resistance tuned. For that reason you should know the general-case condition for parallel-tuned circuits.

Time constant circuits, like the ones in Fig. 3-3, are also very important in the journeyman test. These circuits are not usually shown individually on the CET test as in this illustration but they will very likely be part of a more complete circuit.

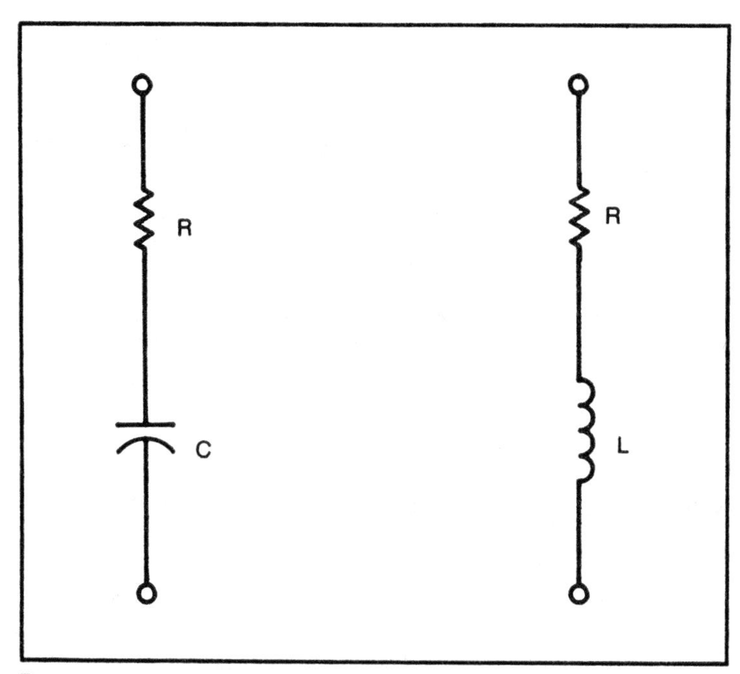

Fig. 3-3. Example of time-constant component combinations.

Typical CET Test Question

In the circuit of Fig. 3-4 when capacitor C2 is switched into the circuit the frequency of the oscillator will

(1) increase.
(2) decrease.

Answer: Choice (2) is correct. When capacitors are combined in parallel the capacitance is increased. Increasing the capacitance will also increase the time (T) for one cycle. Since $f = \frac{1}{T}$ it follows that increasing the time for one cycle will reduce the frequency.

Resistive networks that are used to match impedances are called attenuators or pads. As a general rule, the pads use fixed resistors and attenuators use variable resistors (or, have some

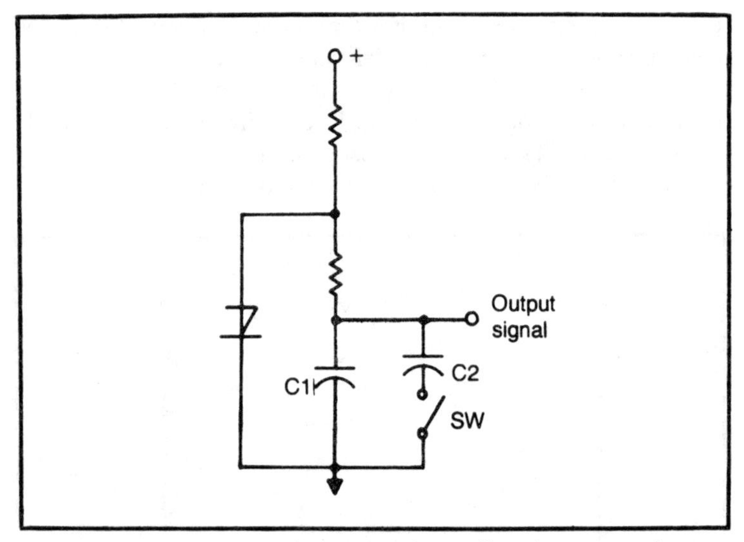

Fig. 3-4. Will the frequency of this relaxation oscillator increase or decrease when switch SW is closed?

provision for changing the resistances in the network). These circuits can be used to match a higher impedance to a lower impedance. In that sense, they perform the same job as a transformer. However, there will be considerably more loss in a pad than there will be in an ideal transformer.

That loss may not be undesirable. For example, you may wish to match a high amplitude signal to a circuit that cannot withstand the high-amplitude input. One good example is, antennas that are mounted in strong signal areas which tend to overdrive the receiver on one channel. The pad can be switched into the circuit to reduce the amplitude of the signal by introducing the loss into the line. If the pad is properly designed, it will not resolve any mismatch of impedances.

One circuit that is very often asked about in the consumer CET test is illustrated in Fig. 3-5. This circuit is sometimes referred to as the *safety circuit*. It is used in transformerless types of receivers and test equipment. Its purpose is to prevent electrocution in the event that the line plug is inserted into the socket in such a way that the chassis becomes hot (above ground). The need for this circuit is eliminated if the ac plug is polarized. However, there is no guarantee that the customer will not at some time replace the plug despite warnings from the manufacturer. So, the safety circuit is still a good idea.

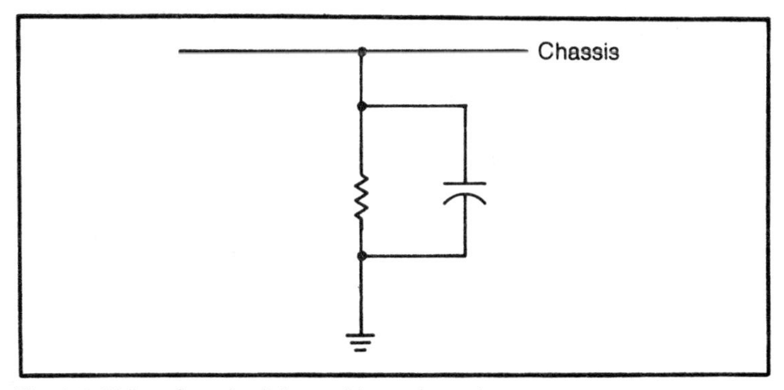

Fig. 3-5. This safety circuit is used in ac-dc equipment.

The problem with this safety circuit is that it is never (or almost never) checked by a technician when he is servicing the equipment. Actually, it should be tested and checked every single time the equipment comes in for repair. When this circuit is defective it does not seriously affect the operation of the receiver so it can be defective and the receiver will still operate satisfactorily.

THREE-TERMINAL AMPLIFYING COMPONENTS

Do not attempt to take the Journeyman-level CET Test unless you are completely familiar with all of the dc operating voltages required on all of the electrodes of the amplifying components shown in Fig. 3-6. It would certainly be a disadvantage to try to troubleshoot electronic equipment if you did not know the typical values—and more important, the polarities—of voltages required for operating these devices.

The polarities in Fig. 3-6 are shown with respect to the charge carrier input terminal (emitter, source, or cathode). However, you may be asked to identify the polarities of voltages between the electrodes. This type of question is demonstrated in the following question.

Typical CET Test Question

In an NPN transistor the base should be

(1) positive with respect to the collector.
(2) negative with respect to the collector.

Answer: Choice (2) is correct. The base must be negative *with respect to the collector*, but it is usually—not always—positive with respect to the common point in the circuit. In other words, the collector is more positive than the base.

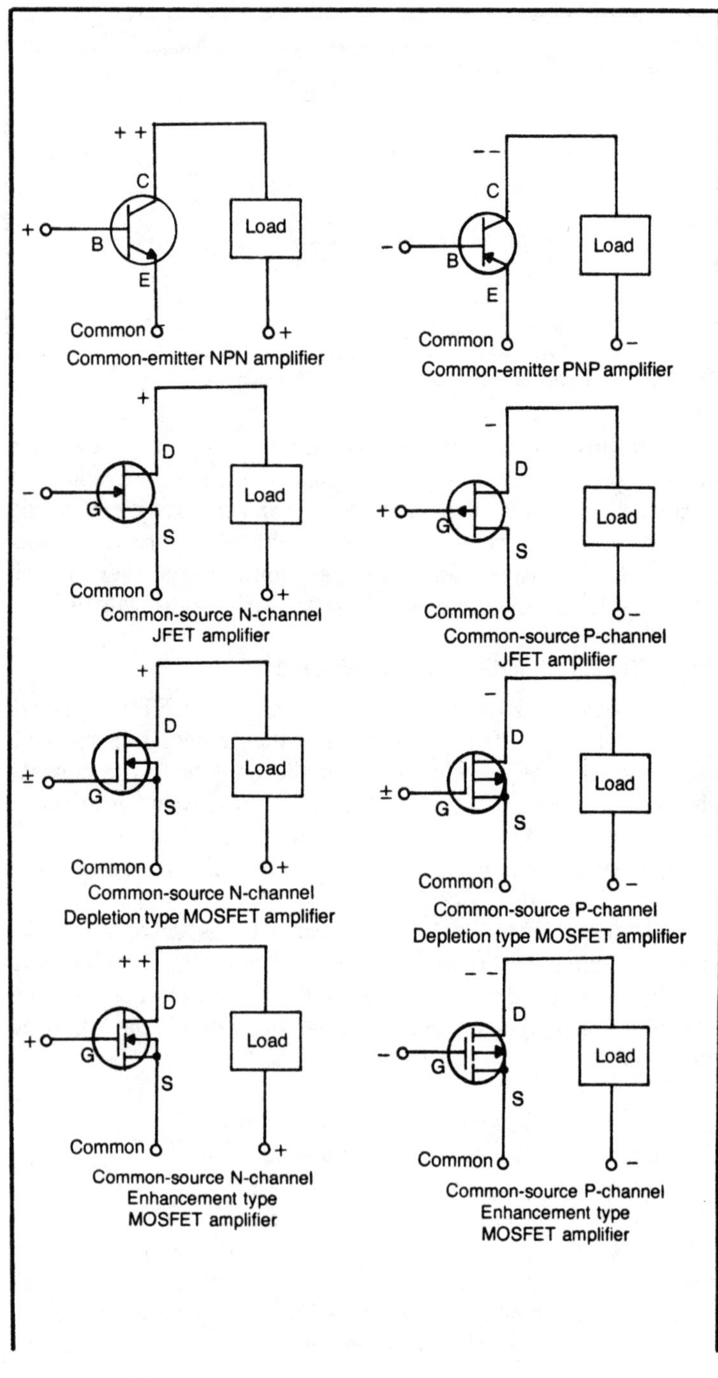

+ +
C
+
B
Load
E
Common
+
Common-emitter NPN amplifier
– –
C
–
B
Load
E
Common
–
Common-emitter PNP amplifier
+
D
–
G
Load
S
Common
+
Common-source N-channel
JFET amplifier
–
D
+
G
Load
S
Common
–
Common-source P-channel
JFET amplifier
+
D
±
G
Load
S
Common
+
Common-source N-channel
Depletion type MOSFET amplifier
–
D
±
G
Load
S
Common
–
Common-source P-channel
Depletion type MOSFET amplifier
+ +
D
+
G
Load
S
Common
+
Common-source N-channel
Enhancement type
MOSFET amplifier
– –
D
–
G
Load
S
Common
–
Common-source P-channel
Enhancement type
MOSFET amplifier

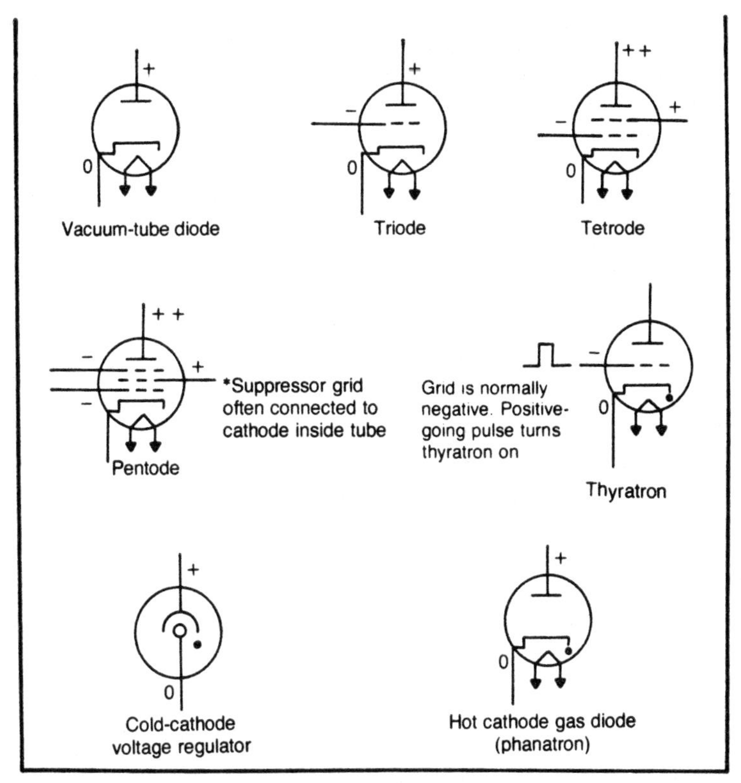

Fig. 3-6. Three-terminal amplifying components and the required dc operating polarities.

There is another reason for studying the illustrations in Fig. 3-6. You must be certain that you understand and have memorized all of the symbols of all of these amplifying components. In the past, technicians have had trouble identifying the depletion versus the enhancement type MOSFETs. They also miss the questions about the difference between the n-channel and p-channel device types. Remember this important fact: *the arrows always point the the N region in electronics symbols!* Therefore, if the arrow is pointing toward the channel it is an n-channel device.

It is presumed that you know the classes of amplification and these will not be reviewed here. Make sure you know them before you go into the test. The classes of interest are: class-A, class-AB1, class-AB2, class-B- and class-C. (There are other classes but they are not discussed or utilized in the CET test.) A few sample questions are given at the end of this chapter to test your knowledge of these classes of operation.

You should also be familiar with the configurations of amplifiers. Speaking in terms of bipolar transistors the three configurations are *common emitter, common base*, and *common collector*. If an FET amplifier is being discussed the classifications are *common source, common gate*, and *common drain*. In vacuum-tube circuits it would be *common cathode, common grid* (or grounded grid), and *cathode follower* (or grounded plate).

Regardless of which amplifying component is being discussed the characteristics of these amplifiers are generally the same. For example, common base circuits have the same general characteristics as common gate and grounded grid amplifiers. They are used in high-frequency circuits where the grounded electrode acts as a shield between the input and the output circuitry.

Common collector (or follower) circuits are utilized where it is desirable to match a high-impedance circuit to a low-impedance circuit. They always have a voltage gain that is less than unity. [$A_v < 1.0$]

Typical CET Test Question

The detector stage of a receiver is generally considered to be a high-impedance circuit because it cannot deliver any significant amount of power to the following stage. Which of the following amplifier configuration would be best for matching the high-impedance detector stage to a low-impedance amplifier:

(1) Common emitter
(2) Common base
(3) Common collector

Answer: Choice (3). The answer is, of course, *common collector*. This circuit, which is also known as an *emitter follower*, has a relatively high input impedance and a lower output impedance.

While I am on this subject I would like to point out that there are special circuits used for increasing the input impedance of bipolar transistor amplifiers. The most important of these is the *bootstrap circuit shown in Fig. 3-7*. When the input signal at "a" goes positive in this circuit the voltage at the top of the emitter resistor (the positive-going voltage) is coupled to "b" by capacitor C. So, the voltage at both sides of the base resistor (R) go positive at the same time. The overall result is that the change in signal voltage across the input resistance does not increase the current through R. If you increase the voltage and it does not increase the current, the resistance must be very large. That's the basic concept of the bootstrap circuit.

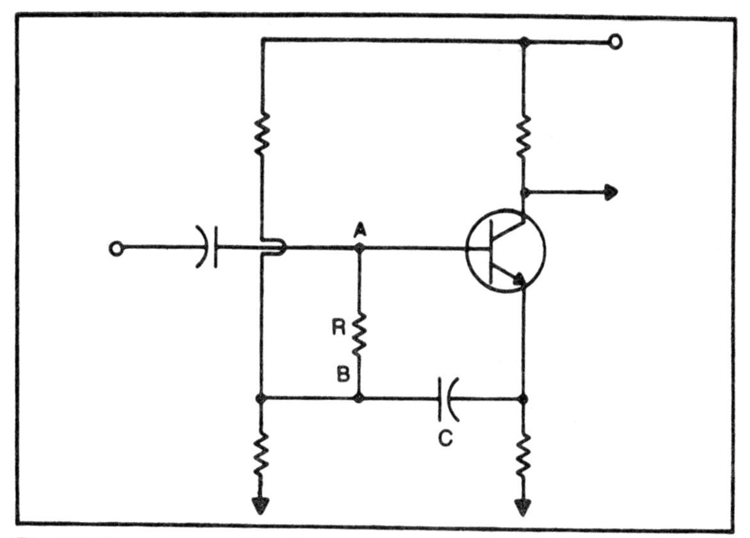

Fig. 3-7. The purpose of this bootstrap arrangement is to increase the input impedances to the amplifier.

Note: The word *bootstrap* is also used for other types of circuits. For example, there is a bootstrap circuit that is used to linearize sweep circuits. You will not be confused between types of bootstrap circuits in the CET test because they will be clearly identified.

You must be thoroughly familiar with the various methods used to bias the amplifying devices. The bias circuits for bipolar transistors are shown in Fig. 3-8.

Not all of these bias circuits have equivalents in field-effect transistor or tube circuitry. For example, vacuum tube, MOSFET, or JFET amplifiers can be biased by putting a resistor in the cathode or source circuit. This is not possible with a bipolar transistor amplifier, although you will find emitter resistors in these circuits. The purpose of those resistors is to stabilize the transistor amplifier against changes in temperature. You are likely to be asked about this emitter resistor. Be sure that you do not ascribe biasing to its purpose.

Figure 3-9 shows some special amplifier configurations that you should understand. These will not be discussed in great detail here, but they will be identified in the following paragraphs.

The Darlington amplifier is sometimes called a *beta-squared amplifier*. The reason is that with this combination the beta of the combined transistor is equal to the product of the beta of each

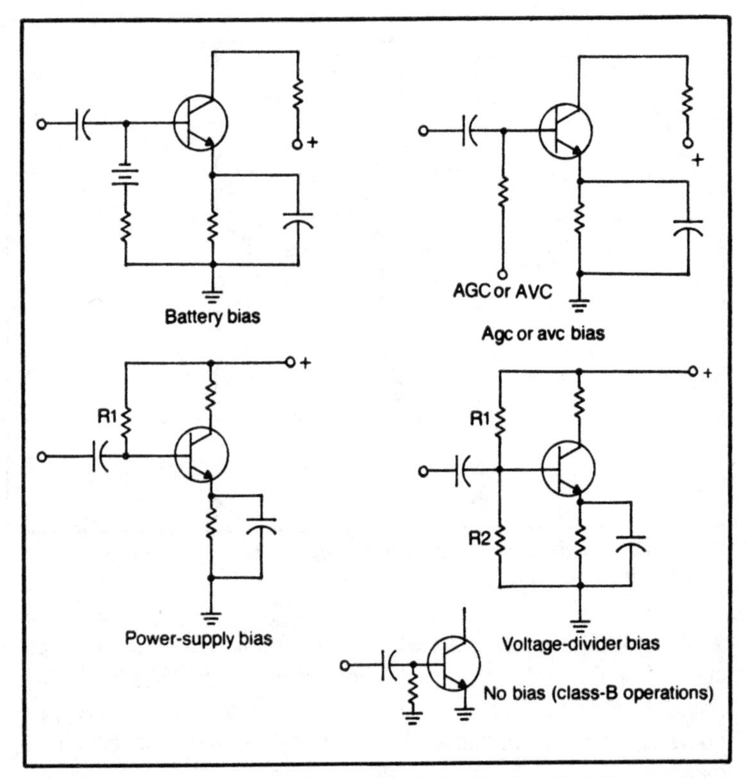

Fig. 3-8. You must know all of the methods for biasing all of the amplifying devices. The example shown here is for bipolar transistors.

transistor. Since Darlingtons are very often connected inside the same case, and they are very closely matched, their betas are the same. Therefore, the beta of the configuration is beta × beta, or, beta squared.

A disadvantage of Darlington circuit is their high internal power dissipation. Their best advantage is that it is possible to get a relatively high gain for a power amplifier configuration.

The totem pole circuit is popular for transistor output circuitry. In this particular example *long-tailed bias* is used. That simply means that a positive supply voltage is used at one end of the amplifier and a negative supply voltage is used at the other end. Long-tail bias permits the load to be operated at or near ground potential when the circuit is idling.

On one-half cycle Q1 conducts, as shown by the solid arrow, and on the next half cycle Q2 conducts, as shown by the dotted arrow. Note that the direction of conduction is opposite on the two half cycles. That means an ac current is flowing through the load resistor. In practice the load resistor for the totem pole may be a speaker.

The complementary amplifiers shown in Fig. 3-9 do not have equivalents in vacuum-tube circuitry. This is a simple circuit combination that permits 180 degree out-of-phase signals to be obtained from a single input. To do this in a tube circuit you need either a phase inverter or a center tapped transformer.

Stacked amplifiers give technicians considerable trouble because the dc circuits for these amplifiers are in series. You can think of the two amplifiers as being a dc voltage divider in addition to being signal amplifiers.

The *cascode amplifier* shown in Fig. 3-9 is shown with vacuum tubes, but equivalent solid state circuitry is also available. The theory of operation for this circuit is that a conventional (common cathode) amplifier is used for the input signal. That signal is direct-coupled to a grounded-grid amplifier. The overall effect is that the amplifier configuration acts like a tetrode but it does not have the accompanying disadvantage of tetrodes—that is, a relatively high partition noise. (Noise will be discussed later in this chapter.)

Differential amplifiers are very important because they are the input amplifier configuration for solid-state operational amplifiers. An important characteristic of differential amplifiers is their common-mode rejection. This means that if you tie the two inputs to the amplifiers together and apply a signal to the junction, there will be (at least, theoretically) no output signal. In other words, there is only an output signal when there is a differential voltage between the two inputs.

You may be asked a question on the CET Test about how differential amplifiers or operational amplifiers reject common-mode signals. You may be asked about reflex circuitry in the Journeyman Consumer CET Test. Remember that an amplifier can be made to amplify two different frequencies at the same time. In Fig. 3-10 the i-f amplifier also serves as the audio voltage amplifier. The advantage of this reflex circuit is that one amplifier can be used for two different signals, so, the circuit is less expensive to amplify.

NOISE IN AMPLIFIERS

There are some types of noise in amplifiers that you should be able to identify. *Partition noise* occurs when a charge carrier moving from the cathode, emitter, or source has two different choices of a direction to go. As an example, when an electron moves from the

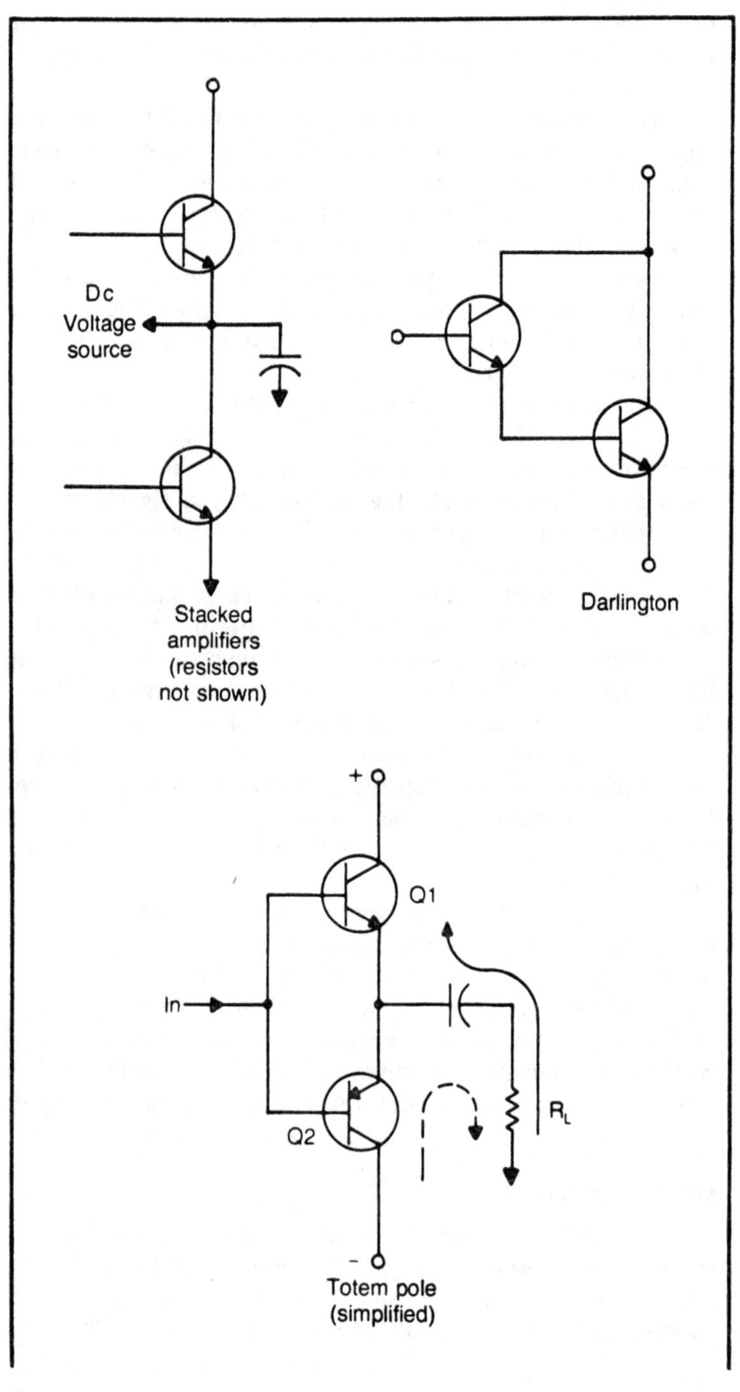
Dc
Voltage
source
Stacked
amplifiers
(resistors
not shown)
Darlington
+
Q1
In
Q2
R_L
-
Totem pole
(simplified)

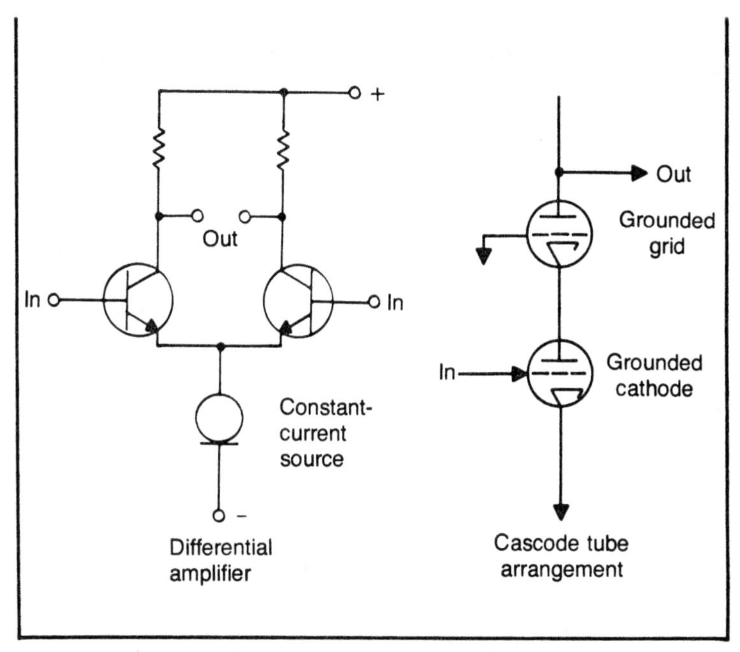

Fig. 3-9. The circuits shown here are simplified. In most cases the resistors have been eliminated. The purpose of the illustration is to show the input and output signal paths for various amplifier configurations.

emitter to the collector in an NPN bipolar transistor it reaches a point where it could go to either the base or to the collector. Both of these electrodes are positive. At any one instant of time there will be a slight difference between the number of charge carriers that go to the base, and the number that go to the collector. This random selection of paths has the effect of producing very small changes in the collector current. These small changes, when flowing through the collector load resistor, will produce undesirable signals called partition noise.

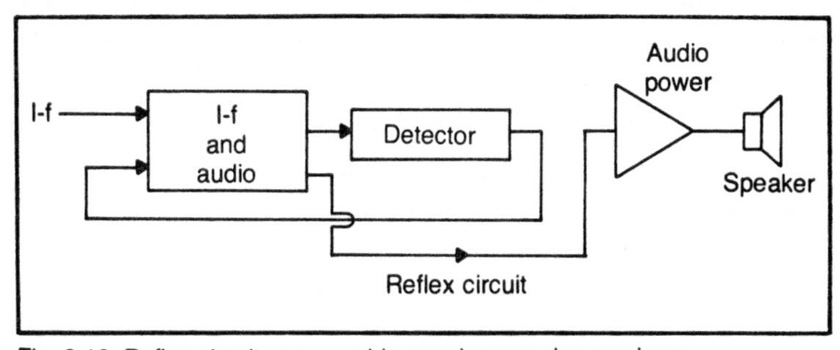

Fig. 3-10. Reflex circuits are used in *very* inexpensive receivers.

Partition noise is a very serious problem in vacuum tubes. Pentodes, of course, will have a much higher partition noise than triodes. That is the reason why the cascode amplifier previously discussed was so very popular in vacuum-tube television receivers. It permitted triodes to be used in the rf section so there was relatively low noise. At the same time, the circuit gave the advantages of tetrodes and pentodes in regard to the isolation of the input and output signals.

MOSFETs have virtually no partition noise since the charge carriers move from the source to the drain without any alternate paths. Therefore MOSFETs are strongly preferred for television high-frequency application such as rf amplifiers and mixers. Partition noise also occurs in bipolar transistors because the number of electrons that cross the emitter-base junction at any single instant is different from the number crossing at the next instant.

Flicker noise occurs in all of the amplifying devices, but it is easiest to understand from the vacuum standpoint. The cathode when emitting electrons, does not emit an identical number of electrons from one moment to the next. This slight change in the number of emitted electrons shows up as small changes in plate current. As with partition noise, these changes will appear as a noise voltage across the plate load resistor. Some flicker noise occurs in MOSFETs because the number of charge carriers injected into the channel varies from instant to instant.

Any time current flows through a resistor there will be a noise signal present due to thermal agitation. Thermal agitation noise is especially prevalent in bipolar and MOSFET devices. In the MOSFET devices the channel between the source and the drain is nothing more than a resistive material. Thermal agitation noise *increases* with an increase in temperature.

Shot noise occurs when the charge carriers disperse when moving from one point to another in an amplifying component. Again, the vacuum tube is used as an example but remember that shot noise occurs in all of the amplifying devices. Suppose you apply a positive-going pulse to the grid of a triode tube. That should release a bunch of electrons, and they should travel from the cathode to plate in a group. Even though this bunch is released simultaneously, they will not arrive at the plate at the same time. You can think of them as being like a handful of lead shot thrown onto a tin roof. They are all thrown at the same time but they do not always arrive at the same time. Therefore the effect is to have a very large number of small electron currents at the destination.

DISTORTION IN AMPLIFIERS

If you connect two different signal frequencies across a perfectly linear resistor the signals do not combine in any way. All you end up with is the two individual frequencies. However, if you take the same two frequencies and introduce them to a nonlinear device *one of the signals will modulate the other.*

In a class-A amplifier there should be no modulation if the amplifier is perfectly linear. Unfortunately, there is no such thing as a perfectly linear amplifier, and therefore, there will *always* be some cross modulation between signals—even in class-A amplifiers. This is known as *intermodulation distortion.* It is a form of nonlinear distortion. The amount of intermodulation distortion can be reduced by operating the amplifier in the most linear portion of its characteristic curve.

If the amplifying device is not biased in the center of its linear characteristic, it is possible for a high-amplitude input signal to be clipped at one end or the other. In other words, either the positive or negative peaks will be clipped. This clipping has the same effect as introducing a great amount of distortion into the signal.

You will remember from your basic theory that a sine wave has absolutely no harmonic content. If you apply a pure sine wave to an amplifier in which clipping occurs there will be a large number of harmonics in the output. For that reason, clipping is a form of *harmonic distortion.*

Sometimes an amplifier is purposely overdriven so that the positive and negative peaks are clipped. The result is a square-wave output signal that has many odd harmonics. Frequency multipliers are often made this way.

No amplifier can produce the same amount of gain to all frequencies in its bandwidth. Any change in the gain at different frequencies is known as frequency distortion. The active components in the output of the amplifier tend to shift the phase of signals at one frequency more than another. This type of distortion is known as *phase-shift distortion.*

SPECIAL RATINGS

Anything that you do to increase the gain of an amplifier will automatically decrease its bandwidth! Conversely, anything that you do to increase its *bandwidth* will decrease its *gain.* I have had many technicians correspond with me about this point. They somehow believe that it is possible to increase both the gain and the bandwidth, but so far, no one has been able to show me how to do it.

The reason this discussion is important here is that one of the ratings of an amplifying component is its gain-bandwidth product. Obviously, a tube, transistor, or FET with a higher gain-bandwidth product can produce a greater gain and greater bandwidth, but there is still a tradeoff within that component for the two limits.

The alpha and beta of a transistor drops off as the frequency increases. Two terms you may encounter are *alpha cutoff* and *beta cutoff*. Those are *frequencies* at which the alpha and beta drop to 70.7 percent of their values at 1 kHz.

A gain-bandwidth product is also a frequency. It is the frequency at which the beta drops to a value of 1.0. The gain-bandwidth product is important for two reasons: it shows that there is a tradeoff between these two parameters; and, it is a rough indication of the high frequency capability of a given transistor or amplifying device.

The bandwidth of an amplifier is often defined as the range of signals between the half power points on its frequency response curve. Another way of saying this is: it is the range of frequencies between the points where the power is down 3 dB. For a response curve that shows voltage versus frequency (instead of power versus frequency) the bandwidth will be between the points where the signal amplitude is down 6 dB.

Remember that a graph showing frequency versus power or frequency versus voltage is called a *bode plot*. Frequency versus phase shift is very often shown on the same bode plot.

The voltage gain of an amplifier is simply the output signal voltage divided by the input signal voltage. For some reason, technicians want to complicate this simple relationship, so, they may miss a very simple question on the voltage gain of a certain amplifier.

Likewise, the power gain is the output power divided by the input power. This simply refers to signal power. It has nothing to do with the total input power of the amplifier. (The total input power would include the operating power that is necessary due to the application of dc operating voltages and currents.)

PROGRAMMED REVIEW

Start with Block number 1. Pick the answer that you feel is correct. If you select choice number 1, go to Block 13. If you select choice number 2, go to Block 15. Proceed as directed. There is only one correct answer for each question.

BLOCK 1

The circuit schematic in this block is for a

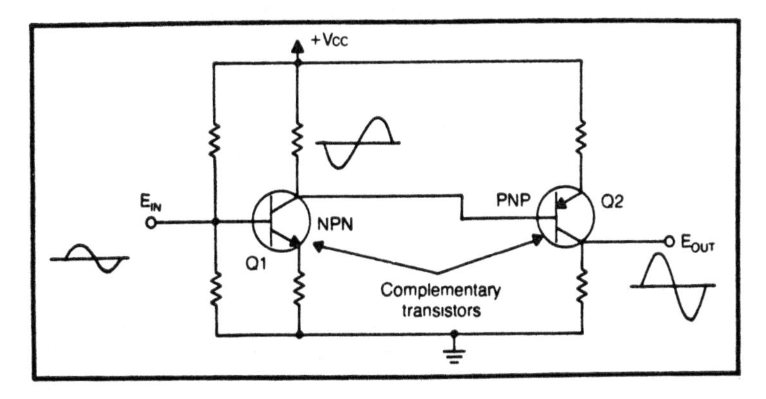

(1) push-pull amplifier. Go to Block 13.
(2) complementary amplifier. Go to Block 15.

BLOCK 2

The correct answer to the question in Block 30 is choice (1). Be sure you know all of the schematic symbols. You can't analyze a circuit from a schematic drawing unless you know what the symbol means.

You might also be asked if the symbol in Block 30 is for an n-channel or p-channel device. The answer is n-channel.

Here is your next question:

For the circuit in this block, the polarity of the supply voltage connected to the POINT "A" should be

(1) positive. Go to Block 12.
(2) negative. Go to Block 17.

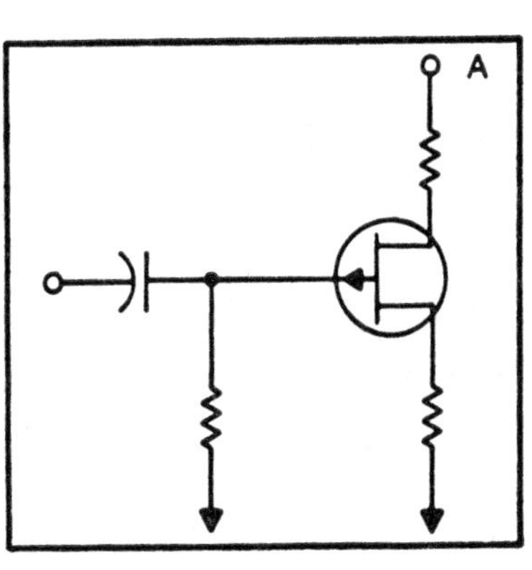

BLOCK 3

Your answer to the question in Block 9 is not correct. Go back and read the question again and select another answer.

BLOCK 4

Your answer to the question in Block 17 is not correct. Go back and read the question again and select another answer.

BLOCK 5

The correct answer to the question in Block 15 is choice (2). In order for a bipolar transistor to be operated class-A it must have forward bias on the base. The circuit shown in Block 15 has no bias, so only the positive half cycles of input signal will be amplified.

Some texts refer to the type of circuit shown in Block 15 as being class-C because a small amount of positive signal is required to overcome the emitter-base junction voltage for class-B operation. Of the choices given, choice (2) is better than choice (1).

Here is your next question:

The purpose of C1 and R2 in the circuit shown in this block is

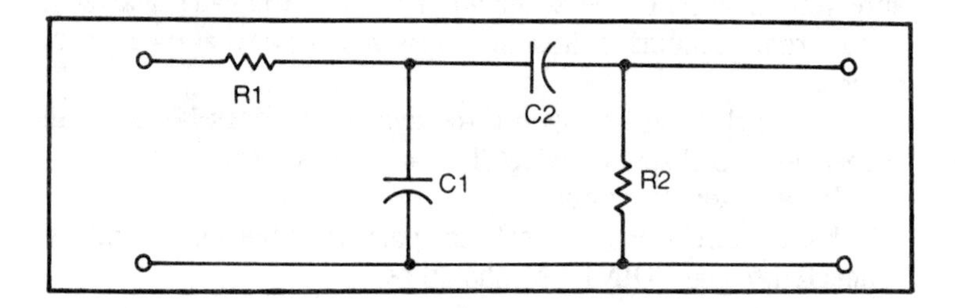

(1) low-frequency compensation. Go to Block 16.
(2) power supply decoupling. Go to Block 21.
(3) cannot tell from the information given. Go to Block 26.

BLOCK 6

Your answer to the question in Block 19 is not correct. Go back and read the question again and select another answer.

BLOCK 7

Your answer to the question in Block 26 is not correct. Go back and read the question again and select another answer.

BLOCK 8

Your answer to the question in Block 15 is not correct. Go back and read the question again and select another answer.

BLOCK 9

The correct answer to the question in Block 19 is choice (1). The triac behaves like back-to-back SCRs. It is a bilateral device, so it is popular in A-C control circuits.

Here is your next question:

Consider the simple SCR circuit shown schematically in this block. If switch SW is closed, then released, the lamp will

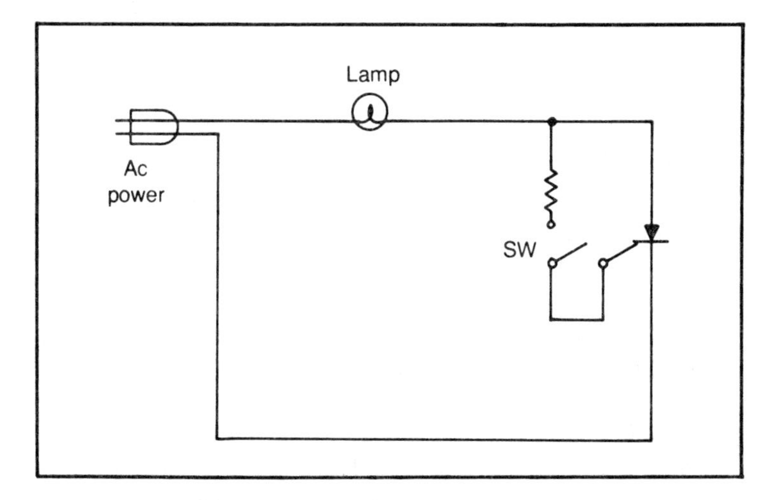

(1) light and stay lit. Go to Block 36.
(2) light only while the switch is closed. Go to Block 30.
(3) not light at all. Go to Block 3.
(4) be On at all times whether or not the switch is closed. Go to Block 20.

BLOCK 10

The correct answer to the question in Block 23 is choice (2). The time constant equals L divided by R. Remember that L is in henries. (100 millihenries = 0.1 henry.)

Here is your next question:

You would expect the amplifier shown in this block to be operated

(1) class-A. Go to Block 29.
(2) class-C. Go to Block 35.

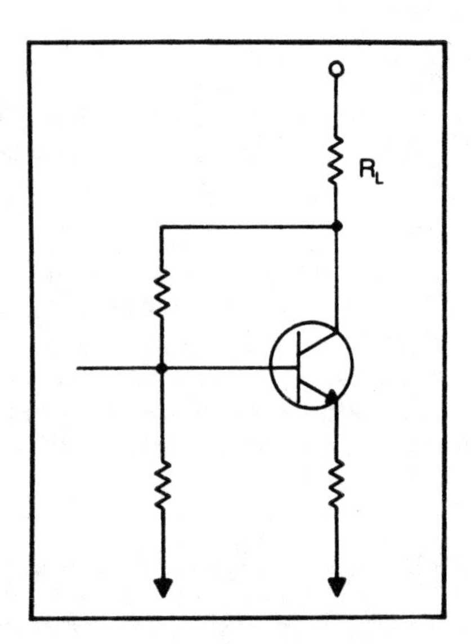

BLOCK 11

Your answer to the question in Block 34 is not correct. Go back and read the question again and select another answer.

BLOCK 12

Your answer to the question in Block 2 is not correct. Go back and read the question again and select another answer.

BLOCK 13

Your answer to the question in Block 1 is not correct. Go back and read the question again and select another answer.

BLOCK 14

Your answer to the question in Block 7 is not correct. Go back and read the question again and select another answer.

BLOCK 15

The correct answer to the question in Block 1 is choice (2). As a general rule, push-pull amplifiers require some type of phase inverter to operate the two amplifying devices. Also, the amplifiers should be matched. In the circuit of Block 1 there is no phase inverter; and, one transistor is an NPN while the other is a PNP.

Here is your next question:

The amplifier shown in this block is operated

(1) class-A. Go to Block 8.
(2) class-B. Go to Block 5.

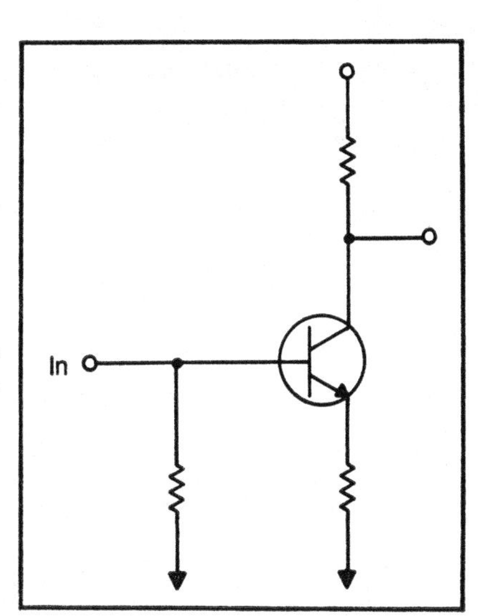

BLOCK 16

Your answer to the question in Block 5 is not correct. Go back and read the question again and select another answer.

BLOCK 17

The correct answer to the question in Block 2 is choice (1). In a

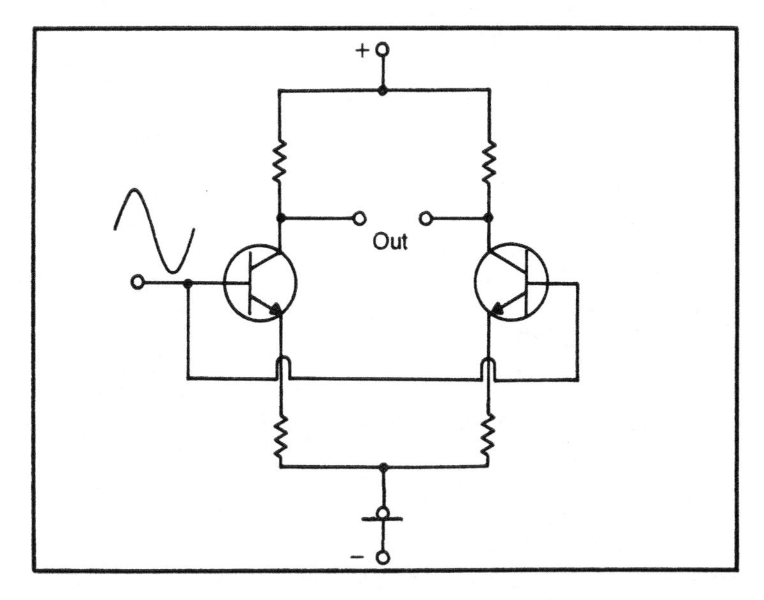

properly-biased, p-channel JFET the drain must be negative with respect to the source. That's another way of saying the source must be positive with respect to the grounded drain.

Here is your next question:

With a pure sine-wave input to the circuit in this block, the output should be

(1) a pure sine wave in phase with the input signal. Go to Block 37.
(2) a pure dc voltage. Go to Block 4.
(3) zero volts. Go to Block 23.

BLOCK 18

Your answer to the question in Block 23 is not correct. Go back and read the question again and select another answer.

BLOCK 19

The correct answer to the question in Block 34 is choice (3). The dc voltmeter should indicate the power supply voltage minus the drop across the load resistor. A sine-wave input *should* produce a sine-wave output signal, and since the full-cycle average of a sine-wave voltage is zero volts, the meter reading should not change. The test shown in Block 34 will check for linearity and clipping.

Here is your next question:

Which of the component symbols represents a thyristor that could be replaced with back-to-back SCRs?

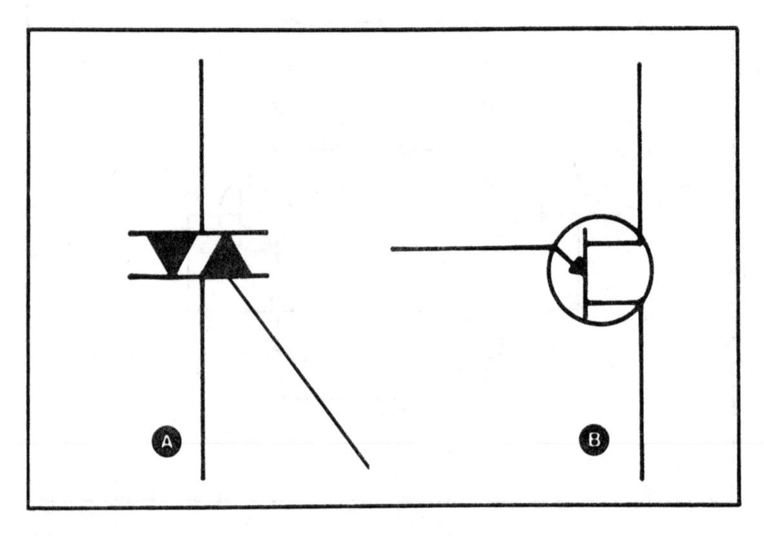

(1) The one shown in A. Go to Block 9.
(2) The one shown in B. Go to Block 6.

BLOCK 20

Your answer to the question in Block 9 is not correct. Go back and read the question again and select another answer.

BLOCK 21

Your answer to the question in Block 5 is not correct. Go back and read the question again and select another answer.

BLOCK 22

The correct answer to the question in Block 32 is choice (2). Since long-tail bias is used, the load is at (or very nearly at) zero volts with respect to common when there is no input signal to the circuit. An input signal causes current to flow back and forth in the load. If the input signal is a sine wave its average for a full cycle is zero volts. That will also be the average of the output signal.

Here is your next question:

You would expect very little partition noise in a

(1) bipolar transistor. Go to Block 27.
(2) MOSFET. Go to Block 34.
(3) pentode tube. Go to Block 31.

BLOCK 23

The correct answer to the question in Block 17 is choice (3). The connection shown in Block 17 tests the differential amplifier for common-mode rejection. If the amplifiers are properly matched they will have equal conduction during all parts of the input signal.

Here is your next question:

What is the time constant of the circuit shown in this block?

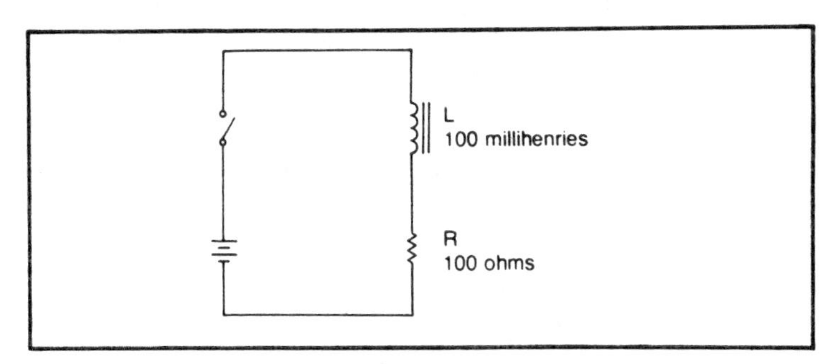

(1) 1 second. Go to Block 18.
(2) 1 millisecond. Go to Block 10.

BLOCK 24

Your answer to the question in Block 30 is not correct. Go back and read the question again and select another answer.

BLOCK 25

Your answer to the question in Block 34 is not correct. Go back and read the question again and select another answer.

BLOCK 26

The correct answer to the question in Block 5 is choice (3). The only way to tell the difference between low-frequency compensation and power supply decoupling is to look at the values of resistance and capacity. These values are not given for the circuit shown in Block 5, so the question cannot be answered.

Here is your next question:

Two different frequencies are fed to a class-A amplifier, and the output signal looks like the one shown in this block. Which of the following statements is correct?

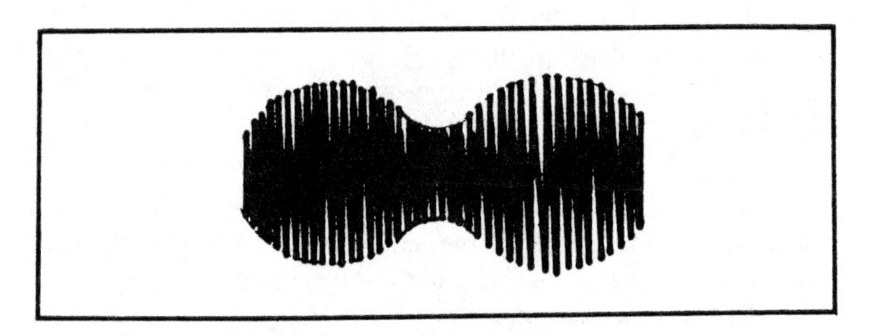

(1) Intermodulation distortion is present. Go to Block 32.
(2) There is nothing wrong. The waveform is correct for class-A amplification. Go to Block 7.

BLOCK 27

Your answer to the question in Block 22 is not correct. Go back and read the question again and select another answer.

BLOCK 28

Your answer to the question in Block 29 is not correct. Go back

and read the question again and select another answer.

BLOCK 29

The correct answer to the question in Block 10 is choice (1). The clues that the amplifier is operating class-A are the voltage-divider forward bias, and, the negative feedback obtained by connecting the voltage divider to the collector rather than to B+.

Here is your next question:

Anything you do to increase the gain of an amplifier will automatically decrease

(1) its operating frequency. Go to Block 28.
(2) its bandwidth. Go to Block 33.

BLOCK 30

The correct answer to the question in Block 9 is choice (2). *If the input power was dc the lamp would light and stay lit when the switch is closed and then opened.*

With an ac input power the lamp will go OFF as soon as the switch is opened. (Actually, the ac power would have to go to zero volts some time during the next half cycle, and at that time the lamp will go OFF. To the human eye, the lamp *appears* to go OFF immediately.)

Here is your next question:

The symbol shown in this block is for

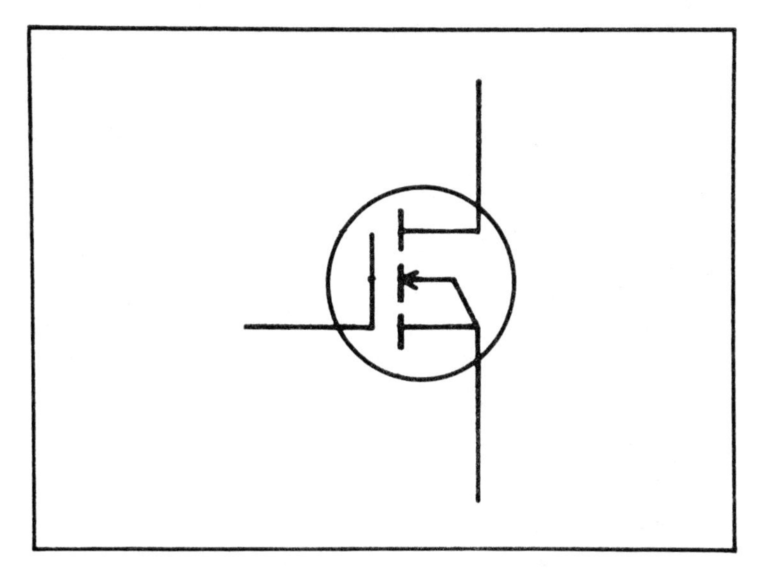

(1) an enhancement-type MOSFET. Go to Block 2.
(2) a depletion-type MOSFET. Go to Block 24.

BLOCK 31

Your answer to the question in Block 22 is not correct. Go back and read the question again and select another answer.

BLOCK 32

The correct answer to the question in Block 26 is choice (1). The procedure described in the question for Block 26 is one that is sometimes used for checking amplifier linearity. Any nonlinearity will result in one signal modulating the other. The resulting waveform is shown in Block 26.

Here is your next question:

If long-tail bias is used in a totem pole configuration, the current flowing through the load—assuming a pure sine-wave input signal—should be

(1) dc. Go to Block 14.
(2) ac. Go to Block 22.

BLOCK 33

The correct answer to the question in Block 29 is choice (2). The operating frequency is determined by the frequency of the input signal. The gain and bandwidth of the amplifier are tradeoffs.

Here is your next question:

Which amplifier configuration is sometimes called *beta squared*?

_______________________________________ Go to Block 38.

BLOCK 34

The correct answer to the question in Block 22 is choice (2). The charge carriers in a MOSFET flow all the way through the channel. There is no choice of current paths (except for the *very* small leakage current in the gate junction) so there is practically no partition noise. This is one reason MOSFETs are preferred for rf amplifiers.

Here is your next question:

Consider the circuit schematic of a class-A amplifier shown in this block. If the 2 kHz sine-wave input signal increases in amplitude, the output (as measured by the dc voltmeter) should

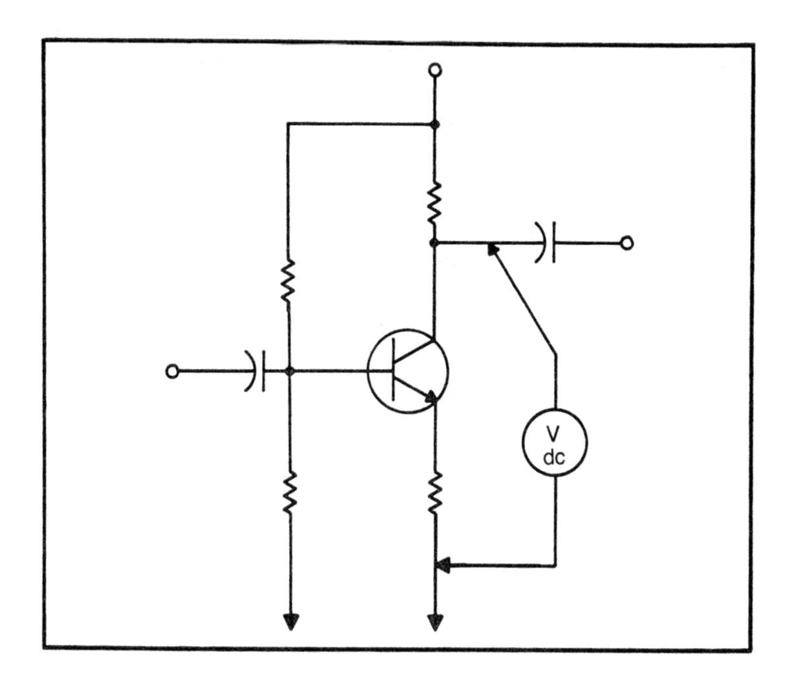

(1) increase. Go to Block 25.
(2) decrease. Go to Block 11.
(3) not change. Go to Block 19.

BLOCK 35

Your answer to the question in Block 10 is not correct. Go back and read the question again and select another answer.

BLOCK 36

Your answer to the question in Block 9 is not correct. Go back and read the question again and select another answer.

BLOCK 37

Your answer to the question in Block 17 is not correct. Go back and read the question again and select another answer.

BLOCK 38

The correct answer to the question in Block 33 is *Darlington*. In a Darlington amplifier the beta of the combination of transistors is equal to the product of the beta of each transistor. If the transistors are closely matched, as in the type where they are both in the same class, the product is: beta × beta = $(beta)^2$.

You have now completed the programmed section.

ADDITIONAL PRACTICE CET TEST QUESTIONS

Some of the questions in this group were used in the past; or, they are contemplated for future use. Answers are given at the end of the chapter.

01. A transistor that does not have a collector is the

(1) unijunction transistor.
(2) junction transistor.
(3) MADT.
(4) PEP transistor.

02. A transistor that may be operated in the enhancement mode is the

(1) MOSFET.
(2) MADT.
(3) PEP.
(4) drift transistor.

03. Another name for a three-layer diode is

(1) a tunnel diode.
(2) a Shockley diode.
(3) a diac.
(4) an LDR.

04. If there is a small amount of negative voltage in the gate of an SCR it will

(1) decrease the leakage current.
(2) prolong the life of the SCR.
(3) decrease the breakover voltage.
(4) increase the breakover voltage.

05. If measurements show that electron current is flowing from the anode to the cathode in a zener diode, the indication is

(1) that the diode is in the circuit backwards.
(2) that the direction of current flow is proper.

06. Which of the following is an application of a tunnel diode?

(1) A detector
(2) An oscillator
(3) A voltage regulator
(4) A high-voltage rectifier

07. Increasing the reverse bias on a varactor diode will

(1) decrease its capacitance.
(2) increase its capacitance.
(3) not affect its capacitance.

08. A reverse bias on a PN junction diode will always destroy a germanium diode if

(1) the reverse bias reaches the zener value.
(2) the reverse bias is pure dc.
(3) the reverse bias is pulsed.

09. Which diode has a negative resistance section on its characteristic curve?

(1) Diac
(2) Breakdown diode
(3) Shockley diode
(4) Esaki diode (also known as a tunnel diode)

10. The base-collector junction of a transistor is normally

(1) always some form of hot carrier construction.
(2) reverse biased.
(3) forward biased.
(4) always passivated.

11. Current gain in a common-emitter transistor circuit is called

(1) the alpha.
(2) the impedance transfer.
(3) the gain-bandwidth product.
(4) the beta.

12. A ratio-detector circuit might use

(1) four-layer diodes.
(2) three-layer diodes.
(3) varactor diodes.
(4) PN junction diodes.

13. A class-A amplifier might be made with

(1) a triac.
(2) a UJT.
(3) a MOSFET.
(4) an SCR.

14. The electrode on a triode that corresponds to the source of a JFET is the

(1) plate.

(2) control grid.
(3) cathode.
(4) screen.

15. Which of the following conducts in either of two directions?

(1) NPN transistors
(2) Triacs
(3) SCRs
(4) Bipolar transistors

16. The polarity of voltage on the gate with respect to the drain of an n-channel JFET is normally

(1) positive.
(2) negative.

17. The frequency at which the common-emitter current gain drops to 70.7 percent of its value at 1 kHz is called

(1) alpha cutoff frequency.
(2) admittance transfer frequency.
(3) gain-bandwidth product.
(4) beta cutoff frequency.

18. Which of the following is nearest to a thyratron in its operation?

(1) MOSFET
(2) SCR
(3) MADT
(4) Diac

19. Which of the following is a type of thyristor?

(1) Tunnel diode
(2) LDR
(3) Junction diode
(4) Diac

20. Which of the following types of diodes is sometimes used as a fast-acting switch?

(1) Diacs
(2) Junction diodes
(3) Tunnel diodes
(4) Zener diodes

21. A P-channel FET should have a gate/cutoff voltage that is

(1) positive with respect to the voltage on the source.
(2) negative with respect to the voltage on the source.

22. Which of the following is NOT a thyristor?

(1) An Esaki diode
(2) An SCR
(3) A triac
(4) A diac

23. The voltage on the collector of an NPN transistor audio amplifier is positive, and the voltage on the base is negative with respect to the emitter. Which of the following is true?

(1) The polarity of the voltage on the collector is wrong.
(2) The polarity of the voltage on the base is wrong.
(3) The polarities of both voltages are wrong.
(4) This is normal.

24. Another name used for a four-layer diode is

(1) a diac.
(2) an Esaki diode.
(3) a hot carrier diode.
(4) a Shockley diode.

25. Which of the following is an example of a breakover diode?

(1) LDR
(2) PN junction diode
(3) Shockley diode
(4) Tunnel diode

26. The operation of a diac is most nearly like the action of

(1) a neon lamp.
(2) a JFET
(3) a thyratron.
(4) a phanatron.

27. To forward bias a semiconductor diode

(1) its anode is made positive with respect to the cathode.
(2) its anode is made negative with respect to its cathode.

28. Which of the following semiconductor devices would be used for obtaining a relatively constant voltage?

(1) A four-layer diode
(2) A diac
(3) A zener diode
(4) A tunnel diode

29. The minority charge carriers in P-type material are

(1) positrons.
(2) neutrons.

(3) electrons.
(4) holes.

30. Another name for Esaki diode is

(1) tunnel diode.
(2) Shockley diode.
(3) diac.
(4) breakdown diode.

31. Charge carriers do not exist in a diode

(1) Barren Straits.
(2) no-man's land.
(3) depletion region.
(4) waste region.

32. An SCR electrode that corresponds to the grid of its equivalent tube device is the

(1) gate.
(2) drain.
(3) emitter.
(4) anode.

33. In a common-base transistor circuit the current gain is called

(1) the alpha.
(2) the impedance transfer.
(3) the beta
(4) the gain-bandwidth product

34. The common-base current gain drops to 70.7 percent of its 1 kHz value at a point called

(1) gain-bandwidth product.
(2) admittance transfer frequency.
(3) beta cutoff frequency.
(4) alpha cutoff frequency.

35. The common-emitter forward-current transfer ratio is unity at a point called

(1) beta cutoff frequency.
(2) alpha cutoff frequency.
(3) gain-bandwidth product.
(4) admittance transfer frequency.

36. The electrode on a JFET that corresponds to the plate of a triode is called

(1) drain.

(2) gate.
(3) base.
(4) source.

37. Which of the following is a bipolar transistor?

(1) NPN
(2) Diac
(3) MOSFET
(4) JFET

38. Which of the following would be grounded in a source-follower?

(1) Drain
(2) Collector
(3) Gate
(4) Source

39. To stop an SCR from conducting

(1) drive the gate voltage negative with a pulse.
(2) make the anode and cathode voltages equal.

40. When a UJT is used as a relaxation oscillator the timing capacitor is connected between

(1) emitter and base.
(2) emitter and collector.
(3) a base and source.
(4) (None of these choices is correct.)

41. As a general rule,

(1) triodes are noisier than pentodes.
(2) tube noise is greater than FET noise.

42. For the circuit in Fig. 3-11, the resonant frequency

(1) can be adjusted with R_C or R_L.
(2) cannot be adjusted by R_C or R_L.

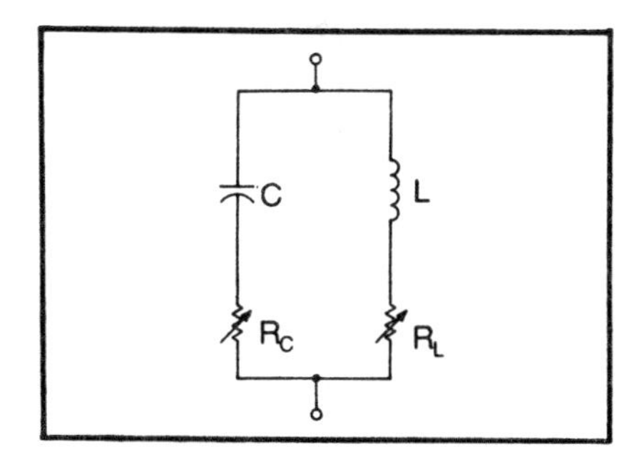

Fig. 3-11. A resonant circuit.

ANSWERS TO PRACTICE TEST

Question Number	Answer Number
01	(1)
02	(1)
03	(3)
04	(4)
05	(2)
06	(2)
07	(1)
08	(1)
09	(4)
10	(2)
11	(4)
12	(4)
13	(3)
14	(3)
15	(2)
16	(2)
17	(4)
18	(2)
19	(4)
20	(3)
21	(1)
22	(1)
23	(2)
24	(4)
25	(3)
26	(1)
27	(1)
28	(3)
29	(3)
30	(1)
31	(3)
32	(1)
33	(1)
34	(4)
35	(3)
36	(1)
37	(1)
38	(1)

Question Number	Answer Number
39	(2)
40	(1)
41	(2)
42	(1)

Antennas and Transmission Lines

One section of the Journeyman-level Consumer CET Test is titled " Antennas and Transmission Lines." This is a test of practical *applications*, and you will not be required to make calculations on dB loss. Nor will you be required to calculate the impedance of specific transmission lines.

There are three categories of questions in this section. One deals with the *transmission lines* and their characteristics. Another deals with *antennas* and their characteristics. The third category deals with *antenna installations*, and specifically, the safety of an installation as it protects against lightning protection.

You (or your company) may not utilize all of the lightning protection described in this chapter. As a technician, you should consult local codes and ordinances before installing an antenna in any area. The procedures described in this chapter represent an accumulation of ideas from Certified Electronics Technicians in all parts of the United States.

TRANSMISSION LINES

Figure 4-1 shows some of the types of the transmission lines used for television installations. By far, the most important of these are the parallel-line transmission type, which is often called *twin lead*, and the *coaxial cable*.

Twin lead is an example of a *balanced line*. That means that at any instant of time the voltage on the two lines is the same with respect to ground. Coaxial cable on the other hand is an example of an *unbalanced line*. It has one conductor (in the center) which carries the signal and an outer conductor which is a braided shield. That outer conductor is normally grounded, and it protects the inner conductor from interference due to stray electromagnetic signals.

The shielded twin lead shown in Fig. 4-1 never became very popular because of the high cost. Another disadvantage is the fact that it is somewhat difficult to work with.

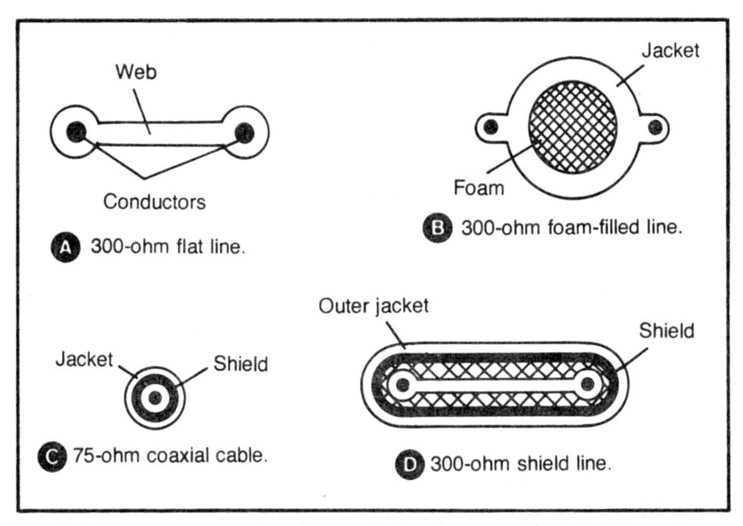

Fig. 4-1. These are examples of transmission lines used for television antenna installations.

An important characteristic of a transmission line is its *characteristic impedance*. (This is sometimes referred to as the *surge impedance* of the line.) That impedance can be calculated from the physical characteristics of the line. Specifically, the diameters of the conductors and the distance between the conductors are very important factors that determine the impedance. The length of the transmission line is *not* a factor that determines its impedance.

Typical CET Test Question

If you cut a 100 foot length of 300-ohm twin lead into two equal sections, what will be the impedance of each 50 foot section?

(1) 0 ohms
(2) 75 ohms
(3) 150 ohms
(4) 300 ohms

Answer: Choice (4), 300 ohms is correct. Cutting the transmission line in half does not change its impedance.

One enterprising technician claimed credit for this question in its original form. That question simply stated that the transmission line was cut in half and what was the impedance of each half? The technician said that the question could not be answered because it was not stated in the test how the line was cut, and he envisioned the line being cut lengthwise. If you get a similar question on the CET test, you can be sure that possibility has now been eliminated.

In comparing coaxial cable with twin lead, remember that coaxial cable is preferred in places where extraneous electromagnetic noise is liable to be present. Also, coaxial cable can be run close to metal, and in fact, it can be run through a metal pipe without affecting its impedance. However, coaxial cable does have a higher loss per foot than twin lead-specifically at the lower television frequencies.

Twin lead is preferred in the lower frequencies (VHF) installations, but at the UHF frequencies it can be quite troublesome. Moisture and dirt collecting on the insulation of this type of line can act as a virtual short circuit at UHF frequencies. Furthermore, twin lead is susceptible to a change in impedance when it is run close to metal objects.

The problem of picking up noise in twin lead can be partially offset by twisting the line. Generally, a twist of one and a half turns per yard is sufficient. (The effectiveness of that procedure has been questioned by some technicians. At the very least you would have to say it doesn't always work.)

The *tubular twin lead* (shown in Fig. 4-1) eliminates some of the problem of loss of transmission energy due to moisture and dirt. The fact that the insulating material between the conductors is in a circular form tends to keep the moisture away from the region between the conductors. However, the greater tendency today is toward the use of single-lead coaxial cable. Most television sets have a provision (in the receiver circuitry) for accepting either 300-ohm or 75-ohm line.

A special transformer called a ***balun*** can be used to match a balanced line to an unbalanced line. Matching can also be accomplished by the use of a resistor network which is usually called a *pad*.

It is important to make sure that the impedances in the transmission line are matched. Otherwise, standing waves will occur on the line. Standing waves are the result of energy being incompletely absorbed at the receiver end. Since the energy is not all absorbed, some of it is returned along the line where it combines with the incident energy coming from the antenna. The result is high-amplitude standing waves along the length of the transmission line.

Standing waves represent a loss of energy, and in the case of twin lead, can result in reradiation of the signal. More important, the standing waves would not occur if all the energy delivered by the antenna and transmission line was absorbed by the receiver. In other words, the *standing waves indicate that there is not an ideal transfer of energy from the antenna to the receiver.*

INSTALLATION OF TRANSMISSION LINES

In the early days of television technicians were using matching stubs made out of 300-ohm transmission lines. These stubs were very effective in reducing the standing wave problem *at some particular frequency.* In those early days there was only one station in the area so the stub did not represent a serious problem. However, a quarter-wave stub at one frequency can be as much as a half-wave or full-wave stub at some other frequency. Therefore, when there is more than one station present, matching stubs should never be used.

In running the transmission line from the antenna to the receiver it is best to avoid long, horizontal runs which can collect moisture and dirt. Also, these horizontal runs act as antennas for horizontally-polarized electromagnetic interference.

Ideally, the transmission line is located as far away as possible from street traffic which may cause interference. Also, it should be run away from buildings that contain rotating machinery. Keep in mind that foam filled types of transmission lines must always be sealed against the possibility of moisture getting in the transmission line. As mentioned before these lines are not very popular today.

If a transmission line is broken at any point between the antenna and receiver there will be two possible symptoms which you should easily recognize. If the antenna is broken completely and does not periodically make connection, the result will be a snowy picture. The reason for the snowy picture is that insufficient signal is being delivered to the receiver. Keep in mind the fact that snow in a picture is not normally picked up by the antenna. Instead, snow is generated almost exclusively by components in the tuner.

The second symptom of a broken transmission line is flashing white streaks in the picture accompanied by static. This is caused by the broken wires in the transmission line momentarily meeting, then breaking apart. This problem becomes very pronounced on windy days.

ANTENNAS

Hertz antennas are used almost exclusively in television receiving systems. A hertz antenna is a half-wave type (see Fig. 4-2). It is mounted horizontally because the television signal is horizontally polarized. (The polarization of a wave is determined by the direction of its electric field. In the case of a television signal, the electric field is horizontal.)

Figure 4-2A shows a simple dipole. It is the simplest example of a hertz antenna. This type is sometimes called a "center-fed hertz" to distinguish it from the "end-fed" type. Note: You will ***not*** be asked about the signal voltage and current distribution on the antenna.

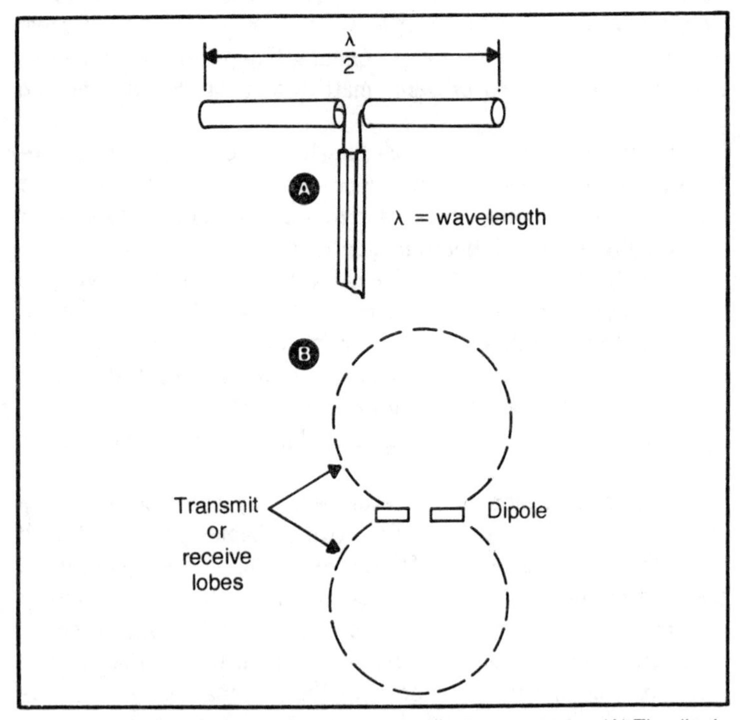

Fig. 4-2. The center-fed dipole is an example of a hertz antenna. (A) The dipole and transmission lines are shown here. Note that the total length of the dipole is one-half wavelength. (B) The directivity of the antenna is shown by the lobes. Adding a reflector will eliminate one of the lobes.

The radiation pattern of the simple dipole—which is the same as the receiving pattern—is shown in Fig. 4-2B. In broadcast AM systems the antenna is vertical and it is a quarter-wavelength long. That type of antenna is referred to as a *Marconi*.

So far, we have talked only about the driven element in the antenna. The driven element is the part that actually receives the signal. A reflector and one or more directors may also be used. You can see these in the illustration of the Yagi antenna in Fig. 4-3. Note that the driven element delivers the signal to the transmission line.

The reflector and directors are called *parasitic elements*. The *directors* receive the signal and reradiate it so that it arrives at the driven element in phase with the original signal. The *reflector* reflects the received antenna back to the driven element. The Yagi antenna of Fig. 4-3 has a very high *gain*, and it is very *directional*.

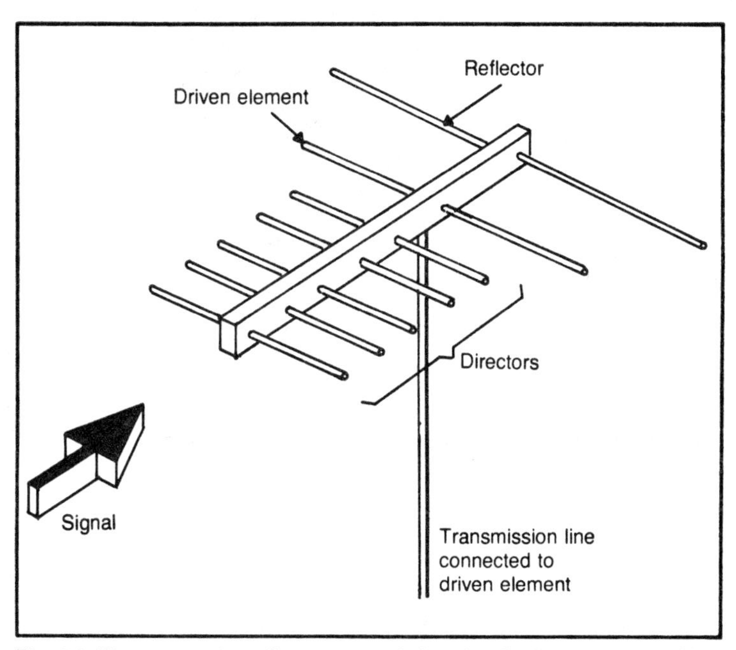

Fig. 4-3. Directors and a reflector are added to the dipole to make this Yagi antenna.

The gain of an antenna is a measure of how well it receives the signal in some particular location compared to how well an *isotropic antenna* would receive the same signal. An isotropic antenna is a mathematical tool. No such antenna actually exists. Specifically, an isotropic antenna is one that receives or transmits equally well in all directions. A half-wave dipole like the one in Fig. 4-2A is normally used for *measurements* of antenna gain. The measurement then compares the signal produced by the antenna in question with the signal received by a dipole in the exact same location. A problem with the Yagi antenna is that it has a very narrow bandwidth.

The *log-periodic* antenna of Fig. 4-4 has the advantage that it can receive a very wide band of frequencies with almost equal flat response. For that reason the log-periodic antenna is preferred for color reception because it does not attenuate the color signals.

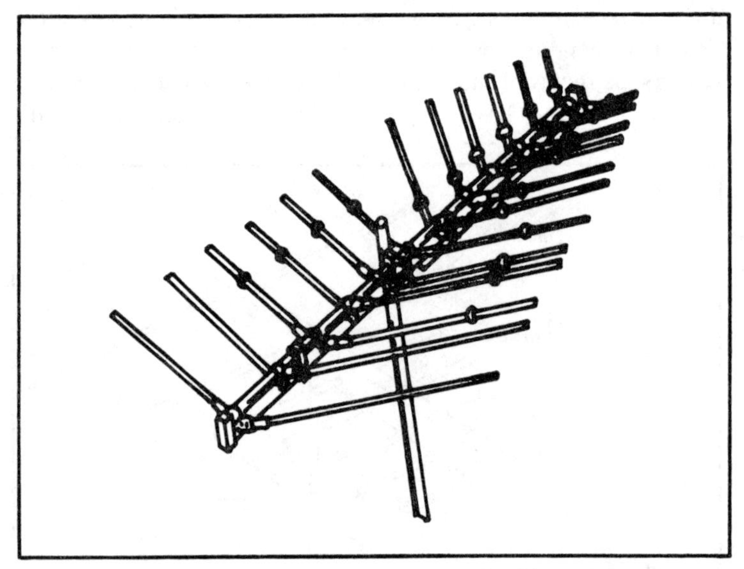

Fig. 4-4. The log-periodic antenna has an extremely wideband response.

A UHF antenna is shown in Fig. 4-5. The small physical size of UHF antennas makes them convenient to work with. However, at UHF frequencies there are problems with reception that do not exist in the VHF range. For example, damp foliage and trees can completely obliterate a UHF signal.

Reflected signals are also more troublesome at the UHF frequencies. If you're in an area where there is ***multi-path distortion*** (ghosts) it may be necessary to use a more directive antenna to eliminate the ghost signals coming from the sides or behind the antenna position. On the television picture the ghost will be on the right side of the desired picture.

It is possible for a misaligned television receiver to produce multiple images which look very much like ghosts. However, in the

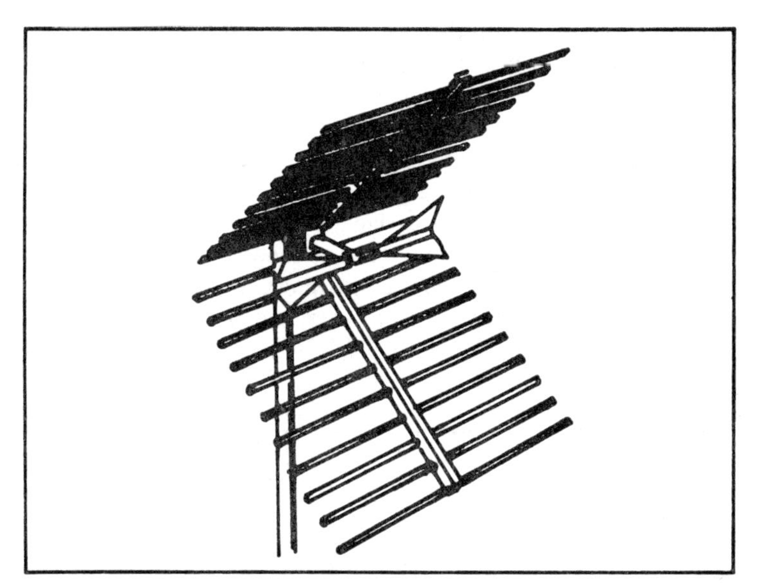

Fig. 4-5. The bow tie is the driven element in this corner reflector. It is an example of a UHF antenna.

case of the misaligned receiver the ghosts are usually tunable. In other words, adjusting the fine tuning control on the receiver can change the position of the ghosts, and also the strength of the ghosts. If the ghost is really coming in from the antenna along the transmission line the fine tuning control will not have a serious effect on it.

It has been proposed that elliptical polarization of the signal could be used to eliminate the ghost problem altogether. Elliptical polarization would mean that the polarization of the television signals periodically rotates. So, the antenna could be located in a position where the reflected signal does not have the proper polarization to induce any appreciable amount of signal into the antenna.

ANTENNA INSTALLATIONS

You may be asked some questions on the CET test about how the customer's home is protected against lightning by a properly installed antenna. Keep in mind the fact that the antenna, if mounted on the roof, is very vulnerable to lightning strikes.

Lightning rods, which are very popular on farmhouses, are grounded conductors that reach above the highest point on the roof. Theoretically, the purpose of the lightning rod is to conduct the lightning safely to ground without permitting it to destroy the house. Of course, with the lightning rod on the house it is more likely to be struck to begin with. However, the properly-grounded lightning rod will keep the lightning on a path outside of the house, and little or no inside damage should result.

When you put an antenna on a roof, you are actually simulating the lightning rod. If the antenna and its mast are not properly grounded, the lightning will strike the antenna but will go *into* the house rather than harmlessly down to the ground.

There are two paths for the lightning to flow. They are shown in Fig. 4-6. One is down the metal mast; and, one other is down the transmission line. Of course, it is also likely that a lightning bolt would take both paths.

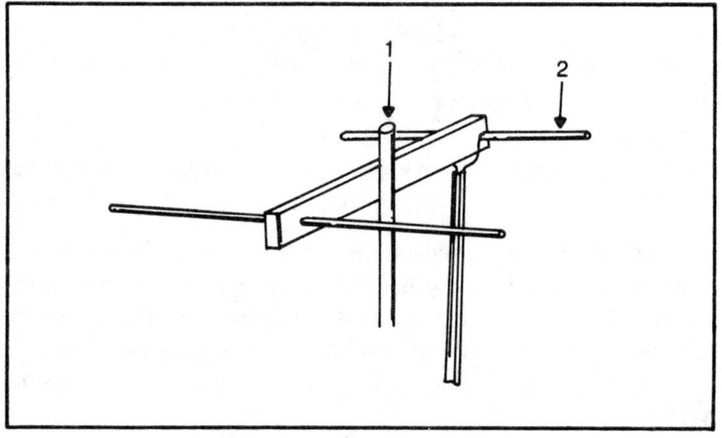

Fig. 4-6. Lightning can strike an antenna on the mast (1) and on the driven element (2).

Consider first the path down the metal mast of the antenna. When the lightning reaches the bottom of the mast it should have a grounded conductor to follow to a safely-grounded point. Immediately, the question arises about grounding to the metal *soil pipe* (or *stand pipe* as it is sometimes called). Can it be safely used as a ground? The answer is a definite NO! In fact, *in many locations there is a local code or ordinance against grounding any kind of antenna to these stand pipes*!

One reason is that plumbers, who are not sympathetic to the cause of antenna installations are liable to put a nonconducting interface between sections of pipe. They don't do this on purpose in order to cause trouble. They just simply aren't thinking about an antenna installation. Therefore, if you ground the mast to these stand pipes, the lightning will enter the house by that path—but it will not go safely to ground. (A likely result is that it will blow the bathroom apart.)

Aluminum ground wire has long been a favorite grounding conductor because it is easy to work with and relatively inexpensive. However, you must remember this about aluminum: *if it is conducting an excessive current the aluminum wire can readily explode*. Obviously, you would never run an aluminum ground conductor inside of the house.

For many years, electricians were wiring homes with aluminum wire, but now there are many local ordinances and codes against it. One reason is that the aluminum does explode when excessive current flows through it. The other reason is that the aluminum wire does not flatten when it is placed in a terminal and affixed by a screw. Therefore, the aluminum wire works loose from the terminals and causes an enormous amount of trouble.

Aluminum wire is still being used for ground wire. It is usually run to a ground rod outside of the home. Some technicians have chosen to connect the aluminum wire to a cold water pipe—specifically one that is used for watering the lawn. *In many locations there is definitely a code against this procedure*!

A very important point to remember about any ground wire is this: do not make right angle turns and complicated paths for the lightning to follow. The reason is that lightning just will not cooperate. Instead of making the right angle turns, it is more likely to enter the house at that point.

The second place where the lightning can cause a problem is the transmission line itself. There have been many cases where the lightning has followed the transmission line into the home and into the receiver. To avoid this, lightning arrestors are sometimes connected to the line outside of the house. This is an important thing to remember about lightning arrestors: **never put the lightning arrestor in the basement, or inside the house in any location**! Make it a rule that you *NEVER* invite lightning into a house.

The arrestors are not foolproof, but in theory they work well. There is an air gap between the transmission line conductor and a grounded conductor. Again, *the grounded conductor should not go in a cold water pipe or some other place which is liable to lead the lightning into the house*. Instead, a ground rod is used for this purpose. It can be the same ground rod as used for the wire coming from the mast.

You should not presume that driving a metal conductor into the ground two or three feet will automatically make a ground connection. If the soil is very sandy, or very dry, there is a possibility that no ground connection will exist even though the rod is inserted deeply into the earth.

To test the ground it is possible to use a 110-volt ac line. Although many technicians use this technique you should understand that *it is very hazardous*. The test is based on the theory that one-half the ac line used for delivering power to a home is grounded. Therefore, an electric drill or a test lamp can be operated by running a wire from the hot side of the line through the drill (or light bulb) back out and to the external ground connection. Figure 4-7 shows the test setup.

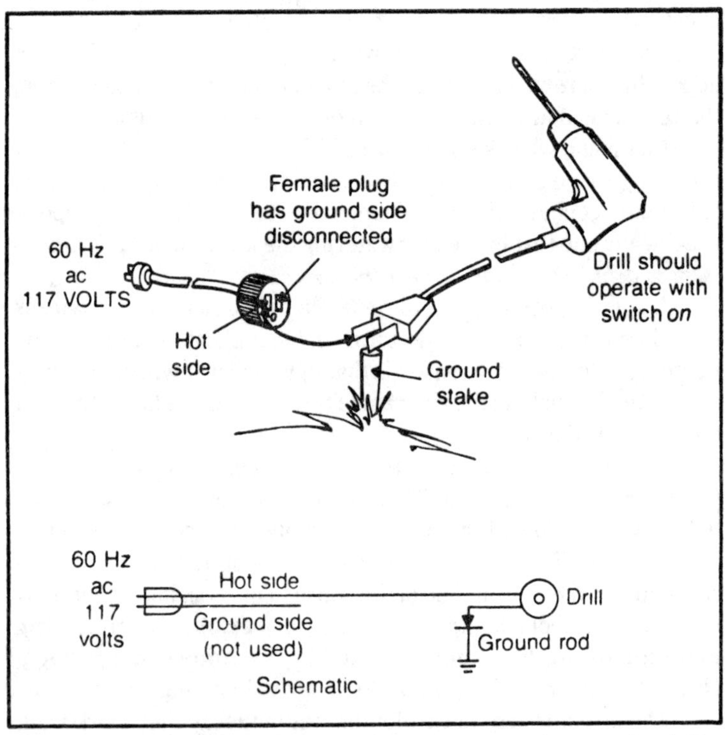

Fig. 4-7. A drill or electric appliance can be used to test the ground rod. The female plug must be modified for this measurement.

Although this procedure is illustrated here, it is not recommended. A much safer procedure is to use a neon test lamp, a test lamp specifically designed for this purpose. There are some common sense questions that you may run across regarding the installation of antennas. Here is an example.

Typical CET Test Question

The antenna of Fig. 4-8 is guyed for mechanical stability. When installing this antenna

(1) the top guy wire should be tightened first.

(2) the bottom guy wire should be tightened first.

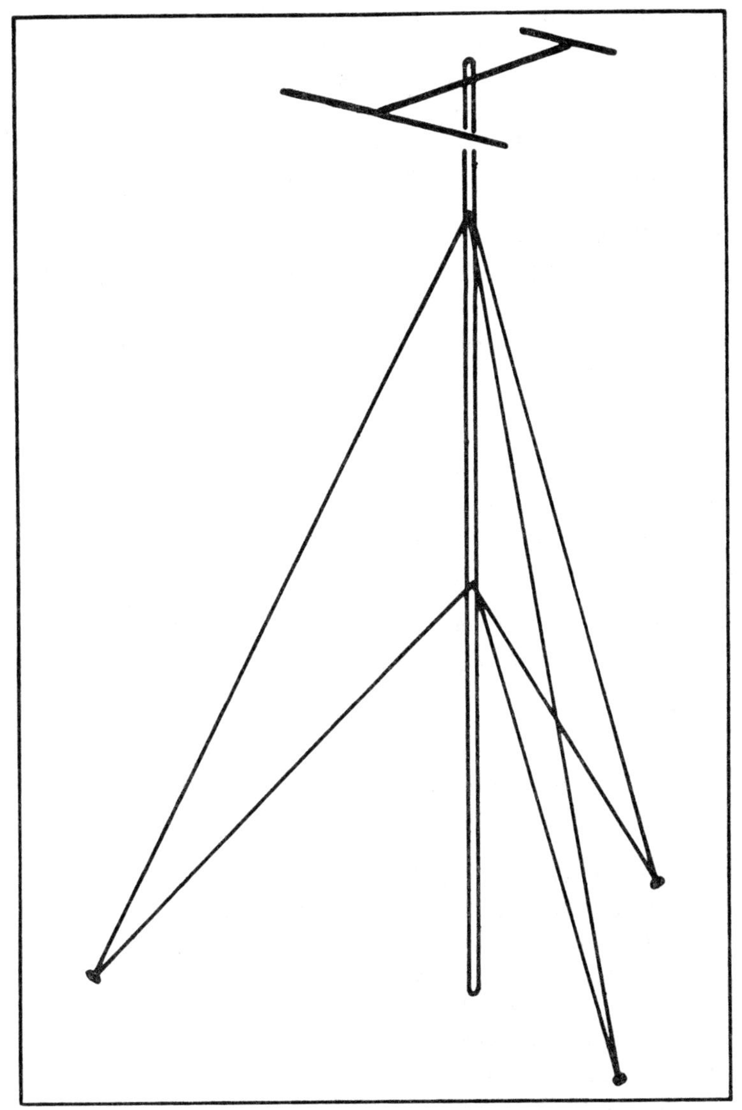

Fig. 4-8. Guide wires are used for tall antenna installations. The one shown here is typical. A 45-degree angle is ideal for the guide wires.

Answer: Common sense should tell you that choice (2) is correct.

If you tighten the top guy wire you will quite possibly put a bend in the mast which you cannot straighten out by tightening the bottom guy wires. Never overlook the fact that when you tighten the guy wires you are pulling the antenna down in a vertical direction toward the roof. If the antenna is not mounted in a strong position, tightening the guy wires can pull the antenna through the roof and, as a matter of fact, through several floors of the house. That is an open invitation to a lawsuit.

Don't overlook the fact that the guy wires can also be a path for lightning to flow. Most technicians prefer to use one of the top guy wires as a path to the ground wire for lightning protection.

PROGRAMMED REVIEW

Start with Block number 1. Pick the answer that you feel is correct. If you select choice number 1, go to Block 13. If you select choice number 2, go to Block 15. Proceed as directed. There is only one correct answer for each question.

BLOCK 1

The television signal is

(1) vertically polarized. Go to Block 13.
(2) horizontally polarized. Go to Block 15.

BLOCK 2

Your answer to the question in Block 15 is not correct. Go back and read the question again and select another answer.

BLOCK 3

Your answer to the question in Block 27 is not correct. Go back and read the question again and select another answer.

Block 4

The correct answer to the question in Block 21 is choice (3). Twisting the line can be overdone, and many technicians have expressed doubt as to whether or not it does any good.

Here is your next question:

If you cut a 100-foot length of 150-ohm transmission into two 50-foot lengths, the impedance of each segment will be

(1) 75 ohms. Go to Block 29.
(2) 150 ohms. Go to Block 5.
(3) 300 ohms. Go to Block 25.

BLOCK 5

The correct answer to the question in Block 4 is choice (2). The impedance is not a function of the length of the line.

Here is your next question:

The driven element for an antenna cut for channel 4 is

(1) longer than one cut for channel 9. Go to Block 11.
(2) shorter than one cut for channel 9. Go to Block 31.

BLOCK 6

Your answer to the question in Block 12 is not correct. Go back and read the question again and select another answer.

BLOCK 7

Your answer to the question in Block 11 is not correct. Go back and read the question again and select another answer.

BLOCK 8

The correct answer to the question in Block 28 is choice (1). To overcome the problem of greater losses at ultra-high frequencies, the transmitters are designed to radiate a greater amount of power. Also, the smaller physical size of the antenna elements makes it possible to design antennas with a higher gain without worrying so much about weight, wind resistance, and vibration.

Remember that UHF signals undergo losses due to weather and vegetation. Due to the fact that these things can change the polarization of the signal, the receiving antenna may not be correctly oriented to get the best reception of the signal.

Here is your next question:

The gain of an antenna is a measure of

(1) how much it increases the strength of the received signal. Go to Block 16.
(2) how much signal it receives compared to how much signal an isotropic signal would receive in the same position. Go to Block 24.

BLOCK 9

The correct answer to the question in Block 23 is choice (1). When standing waves appear on the transmission line it means that not all of the signal energy is being used by the receiver. Instead, some energy is reflected back along the line where it combines with the incoming signal to form the standing waves. Ideally, ***all*** of the signal from the antenna should be used by the receiver.

Here is your next question:

When a transmission line is connected to a simple half-wave dipole, an ohmmeter reading on the receiver end of the line should be

(1) infinity. Go to Block 18.
(2) zero ohms. Go to Block 20.

BLOCK 10

Your answer to the question in Block 21 is not correct. Go back and read the question again and select another answer.

BLOCK 11

The correct answer to the question in Block 5 is choice (1). Antennas are cut to ***wavelength***. The higher the frequency, the shorter the wavelength.

If you measured an antenna cut to channel 4 you would find it to be slightly longer than one-half the calculated wavelength. The reason is that electric signals travel somewhat slower in the antenna metal conductor than in the air. The wavelength equation for free space signals is:

$$\lambda = \frac{300,000,000}{f} \text{ meters}$$

(The frequency must be given in hertz.)

Here is your next question:

Which of the following antenna types is better for wideband reception?

(1) Log periodic. Go to Block 23.
(2) Yagi. Go to Block 7.

BLOCK 12

The correct answer to the question in Block 24 is choice (2).

Here is your next question:

If the transmission line for an installation must run close to metal pipes, it would be better to use

(1) twin-lead 300-ohm transmission line. Go to Block 6.
(2) 75-ohm coaxial cable. Go to Block 21.

BLOCK 13

Your answer to the question in Block 1 is not correct. Go back and read the question again and select another answer.

BLOCK 14

Your answer to the question in Block 18 is not correct. Go back and read the question again and select another answer.

BLOCK 15

The correct answer to the question in Block 1 is choice (2). The advantages of the horizontally-polarized VHF television signal include:

- Antennas designed for horizontal polarization automatically reject man-made noise signals. The reason for this is that man-made noise signals are often vertically polarized.
- Horizontally-polarized waves provide better spacewave coverage.
- It is easier to make the antennas very directive when they are designed for horizontally-polarized waves.
- Horizontally-polarized waves follow the ground better. In other words, the ground waves have less loss than vertically-polarized ground waves.

Here is your next question:

In a horizontally-polarized wave the

(1) electric field is horizontal. Go to Block 28.
(2) electromagnetic field is horizontal. Go to Block 2.

BLOCK 16

Your answer to the question in Block 8 is not correct. Go back and read the question again and select another answer.

BLOCK 17

Your answer to the question in Block 21 is not correct. Go back and read the question again and select another answer.

BLOCK 18

The correct answer to the question in Block 9 is choice (1). This answer assumes that the transmission line is *not* connected to the receiver.

Here is your next question:

A likely cause of white flashes in the picture and accompanying static is

(1) incorrect antenna orientation. Go to Block 14.
(2) a broken lead in the transmission line. Go to Block 27.

BLOCK 19

Your answer to the question in Block 27 is not correct. Go back and read the question again and select another answer.

BLOCK 20

Your answer to the question in Block 9 is not correct. Go back and read the question again and select another answer.

BLOCK 21

The correct answer to the question in Block 12 is choice (2).

Here is your next question:

The reason for twisting 300-ohm twin-lead transmission line is

(1) to prevent standing waves. Go to Block 10.
(2) to increase the line impedance. Go to Block 17.
(3) to cancel noise signals. Go to Block 4.

BLOCK 22

Your answer to the question in Block 23 is not correct. Go back and read the question again and select another answer.

BLOCK 23

The correct answer to the question in Block 11 is choice (1). The most important feature of a log-periodic antenna is its broadband characteristic. The most important features of a Yagi antenna are high gain and high directivity.

Here is your next question:

A 75-ohm coaxial cable delivers a signal to a receiver with a 300-ohm input impedance. The result is

(1) standing waves. Go to Block 9.
(2) a step up in the signal level. Go to Block 22.

BLOCK 24

The correct answer to the question in Block 8 is choice (2). An isotropic antenna receives signals equally well from all directions.

In real life, no such antenna exists. However, it is possible—using a complicated mathematics procedure—to calculate how much signal strength would be received from an isotropic antenna at the position of interest.

Simple diodes are used as the standard receiving antenna in field strength *measurements*, since it is not possible to obtain an actual isotropic antenna. For this measurement, the signal received by the antenna being measured is compared with the signal received by the dipole in the same position. It is customary to express the *dB gain* of the antenna being measured. This shows (indirectly) how many times better the antenna being measured is, compared to the dipole, for receiving a given signal.

Here is your next question:

A balun is

(1) a transformer designed to step up the voltage on a balanced transmission line. Go to Block 30.
(2) used to connect a signal from a balanced transmission line to an unbalanced transmission line. Go to Block 12.

BLOCK 25

Your answer to the question in Block 4 is not correct. Go back and read the question again and select another answer.

BLOCK 26

Your answer to the question in Block 28 is not correct. Go back and read the question again and select another answer.

BLOCK 27

The correct answer to the question in Block 18 is choice (2). The white flashes occur when the antenna line momentarily breaks and then makes. If the antenna is not correctly oriented (choice 1) the result will be a weak picture and (usually) snow.

Here is your next question:

Which of the following is the cause of snow or colored confetti on a television screen?

(1) It is caused by noise generated in the receiver front end. Go to Block 32.
(2) It is caused by noise signals that are picked up by the antenna. Go to Block 3.
(3) It is produced in the video amplifier stage. Go to Block 19.

BLOCK 28

The correct answer to the question in Block 15 is choice (1). Remember that the electric field is in the same plane as the driven element of the antenna.

Here is your next question:

The UHF television signals

(1) have greater losses than the VHF signals. Go to Block 8.
(2) can be transmitted with much lower losses than VHF signals. Go to Block 26.

BLOCK 29

Your answer to the question in Block 4 is not correct. Go back and read the question again and select another answer.

BLOCK 30

Your answer to the question in Block 24 is not correct. Go back and read the question again and select another answer.

BLOCK 31

Your answer to the question in Block 5 is not correct. Go back and read the question again and select another answer.

BLOCK 32

The correct answer to the question in Block 27 is choice (1). Virtually all of the noise generated in a receiver originates in the tuner. Snow and colored confetti are noise signals as they appear on the screen.

Here is your next question:

The antenna element that delivers the signal to the transmission line is called the *driven element*. The reflector and the directors are called the ______________ elements. Go to Block 33.

BLOCK 33

They are called *parasitic elements*.

You have now completed the programmed section.

ADDITIONAL PRACTICE CET TEST QUESTIONS

Some of the questions in this group were used in the past; or, they are contemplated for future use. Answers are given at the end of the chapter.

01. Which of the following is true regarding VHF television signals?

(1) They are not polarized.
(2) They are horizontally polarized.
(3) They are vertically polarized.

02. A horizontally polarized wave has

(1) its electric field in a horizontal direction—that is, parallel to the surface of the earth.
(2) its magnetic field in a horizontal direction—that is, parallel to the surface of the earth.

03. Which spectrum of frequencies is between 30 and 300 MHz?

(1) VHF
(2) HF
(3) UHF
(4) SHF

04. Maximum power will be transferred between the antenna and the receiver if

(1) the transmission line impedance matches only the antenna impedance.
(2) care is taken that the transmission line impedance does not match the antenna or receiver impedance.
(3) the transmission line impedance matches only the receiver impedance.
(4) the transmission line matches both the antenna and the receiver impedance.

05. Which of the following is *not* true regarding an impedance matching?

(1) A mismatch causes less than the maximum power to be delivered to the receiver.
(2) A mismatch causes a ghost image.
(3) There should be a certain amount of mismatch in strong signal areas for best reception.

06. Corner reflectors are not used with VHF television antennas because

(1) their gain is too low.
(2) their impedance is too high.
(3) they are much too bulky and heavy.

07. Which of the following is the proper spacing between stacked antennas?

(1) a half-wavelength.
(2) a quarter wavelength.
(3) one wavelength.
(4) any spacing that is convenient.

08. Which of the following is *not* an advantage of stacking antennas?

(1) A narrower vertical directivity. (This makes it possible for the antenna to reject the reflected ground waves.)
(2) An increase gain over a single antenna.
(3) An increase in impedance.

09. The physical length of a television antenna is

(1) about 95% greater than its electrical length.
(2) about 5% of its electrical length.
(3) about 95% of its electrical length.
(4) about 5% greater than its electrical length.

10. Which of the following is the impedance of a half-wave center-fed dipole antenna?

(1) 300 ohms
(2) infinite
(3) 72 ohms
(4) 200 ohms

11. A horizontal directional pattern of a simple dipole is

(1) omnidirectional.
(2) unidirectional.
(3) bidirectional.

12. Which of the following is a simple half-wave dipole antenna?

(1) Yagi
(2) Marconi
(3) Zepp-fed
(4) Hertz

13. The line-of-sight distance of a television signal would not be increased by

(1) increasing the length of the driven element on the receiving antenna.
(2) raising the height of the transmitting antenna.
(3) raising the height of the receiving antenna.

14. For a line-of-sight transmitted signal, the cutoff point is not actually at the horizon because the signal travels

(1) beyond the horizon by some 50% before the signal strength begins to drop rapidly.
(2) beyond the horizon by some 15% before the signal strength begins to drop rapidly.

15. Which of the following is ***not*** an advantage of horizontally polarized waves for television reception?

(1) Less ground-wave loss.
(2) Signals are not expensive to transmit.
(3) Vertically polarized-noise signals do not cause as much interference.

16. If a guy wire is rated 8 × 10, it would be

(1) 8′ long and 10″ wide.
(2) made of number 10 wire with 8 strands.
(3) made of number 8 wire with 10 strands.
(4) None of these choices is correct.

17. In reference to twin-lead transmission line for television installations,

(1) it has a higher loss than coaxial cable.
(2) attenuation losses are increased when moisture collects on the surface.
(3) it is designed primarily for outside use.

18. How does an antenna-mounted preamplifier normally get the required dc operating power?

(1) Through the transmission line that delivers the signal to the set.
(2) Through a separate line.
(3) With lithium-air batteries.
(4) By nuclear power packs.

19. When an ohmmeter is connected across a transmission line that goes to a dipole antenna (not a folded dipole) it should show

(1) 300 ohms.
(2) about zero ohms.
(3) infinity.
(4) None of these answers is correct.

20. A folded dipole has an impedance of approximately

(1) 450 ohms.
(2) 300 ohms.

(3) 72 ohms.
(4) None of these answers is correct.

21. Which of the following would correctly match the impedance of a simple dipole?

(1) A 300-ohm twin lead.
(2) A 72-ohm coaxial cable.

22. Which of the following antennas could be used for the entire range of TV frequencies from channel 2 to channel 83?

(1) A log-periodic antenna.
(2) A bow tie antenna with a corner reflector.
(3) A folded-dipole stacked array.
(4) A stacked conical antenna.
(5) A stacked Yagi antenna.

23. In comparing a folded dipole with a simple dipole which of the following is *not* true?

(1) The folded dipole is more rigid in structure.
(2) The folded dipole has a higher impedance.
(3) The folded dipole is more directive.

24. The reflector and the directors on a Yagi antenna are called

(1) vestigial elements.
(2) parasitic elements.
(3) restored elements.
(4) driven elements.

25. A negative dB gain means there is

(1) an impossible situation.
(2) an improvement in signal strength.
(3) a negative resistance in the circuit.
(4) a loss in the circuit.

26. Which of the following is the best location for a television preamplifier?

(1) As close to the receiver as possible.
(2) As close to the antenna as possible.

27. The noise susceptibility in an AM system

(1) cannot be measured.
(2) varies directly as the bandwidth.
(3) varies inversely as the bandwidth.
(4) is not related to bandwidth.

28. The longest element of a Yagi antenna is the

(1) driven element.
(2) director.
(3) reflector.

29. Which of the following signal frequencies is not reflected from the ionosphere?

(1) 88 MHz
(2) 17 MHz

30. Twisting a 300-ohm twin-lead transmission line

(1) lowers the line impedance.
(2) increases the line inductance.
(3) reduces the line capacitance.
(4) reduces the amount of noise pickup by the line.

ANSWERS TO PRACTICE TEST

Question Number	Answer Number
01	(2)
02	(1)
03	(1)
04	(4)
05	(3)
06	(3)
07	(1)
08	(3)
09	(3)
10	(3)
11	(3)
12	(4)
13	(1)
14	(2)
15	(2)
16	(3)
17	(2)
18	(1)
19	(3)
20	(2)
21	(2)
22	(1)
23	(3)
24	(2)
25	(4)
26	(2)
27	(2)
28	(3)
29	(1)
30	(4)

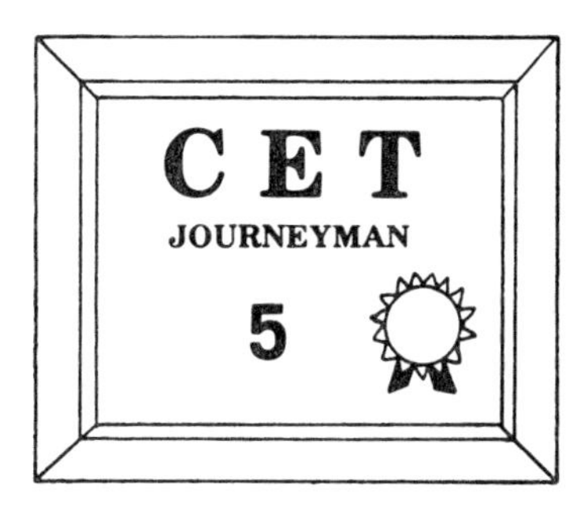

Digital Circuits in Consumer Products

An understanding of digital circuits has become so important for a television technician that 25 questions in the Consumer Journeyman CET Test are devoted to digital logic and microprocessor circuitry. There are a few questions on Boolean algebra and conversion from one numbering system (such as hexadecimal) to another (such as binary). However, for the most part the subject matter in this chapter (and in the corresponding section in the CET Test) deals with the hardware that is associated with digital and microprocessor circuitry.

GATE SYMBOLS AND OTHER IDENTIFICATION

You should not attempt to take the consumer test unless you have a very good understanding of the basic logic gates. Specifically, you should know the *truth table* for each gate, the *symbols* used for the gates, the *Boolean equation* for each gate, and you should be able to recognize *simple circuitry* that represents the basic gates. All of those things will be reviewed in this chapter.

There are seven gates that you *must* understand. They are: AND, OR, NOT, NAND, NOR, EXCLUSIVE OR, AND EXCLUSIVE NOR. The EXCLUSIVE NOR is often called a *logic comparator*. The truth tables for these gates are shown in Table 5-1. Make sure that you can identify the truth tables, and the gates that go with each one.

The truth tables given in Table 5-1 are for two-input logic

Table 5-1. Truth Tables for Basic Gates.

AND

A	B	L
0	0	0
0	1	0
1	0	0
1	1	1

OR Also Called Inclusive OR

A	B	L
0	0	0
0	1	1
1	0	1
1	1	1

NOT

A	L
0	1
1	0

NAND

A	B	L
0	0	1
0	1	1
1	0	1
1	1	0

NOR

A	B	L
0	0	1
0	1	0
1	0	0
1	1	0

EXCLUSIVE OR

A	B	L
0	0	0
0	1	1
1	0	1
1	1	0

EXCLUSIVE NOR (Logic Comparator)

A	B	L
0	0	1
0	1	0
1	0	0
1	1	1

gates. From these you should be able to identify truth tables for three-input, four-input, or any number of inputs. For example, look at the truth table for the AND gate. Note that the only way to get an output of logic 1 is to have all of the inputs at a logic 1 level. So, if it were a three-input AND gate, all three inputs would have to be a logic 1 in order to get a 1 output.

Note also that the only way you can get a logic 1 output from the OR gate is to have all the inputs at logic 0. So, if it is four-input OR gate, all four inputs would have to be at logic 0 in order to get a 0 output. This same reasoning can be applied to the other gates in Table 5-1.

Table 5-2 gives the Boolean equation for each of the basic gates. Observe that there are different equations used to represent the same gates. In the case of the AND gate there are three different ways to express A AND B = L. They are:

$$A \times B = L$$
$$A \cdot B = L$$
$$AB = L$$

You will see the Boolean equation in any of these forms. In the case of the NAND gate, only one version is shown. However, an overbar could be placed on any of the three expressions for AND to obtain a NAND. The examples are shown here:

$$\overline{A \times B} = L$$
$$\overline{A \cdot B} = L$$
$$\overline{AB} = L$$

Observe that the EXCLUSIVE OR gate is represented two different ways with the Boolean equations.

There doesn't seem to be any shortcut in method of memorizing these basic Boolean expressions. If you have worked with them for a short time, you will have no difficulty recognizing any one of them. However, if this subject is fairly new to you, you might try putting them on flash cards to help with the memorization. Table 5-3 shows gate symbols and simple circuits.

IMPORTANT RULES FOR OVERBARS

There are a few rules for overbars that you should also know

Table 5-2. Boolean Equations for Logic Gates.

Gate	Equation
AND	$A \times B = L$ $A \cdot B = L$ $AB = L$
OR Also Called Inclusive OR	$A + B = L$
NOT	$\bar{A} = L$ (or, $A = \bar{L}$)
NAND	$\overline{AB} = L$
NOR	$\overline{A + B} = L$
EXCLUSIVE OR	$A \oplus B = L$ (or, $\bar{A}B + A\bar{B} = L$)
LOGIC COMPARATOR	$\bar{A}\bar{B} + AB = L$

Table 5-3. Logic Gate Symbols and Simple Circuits.

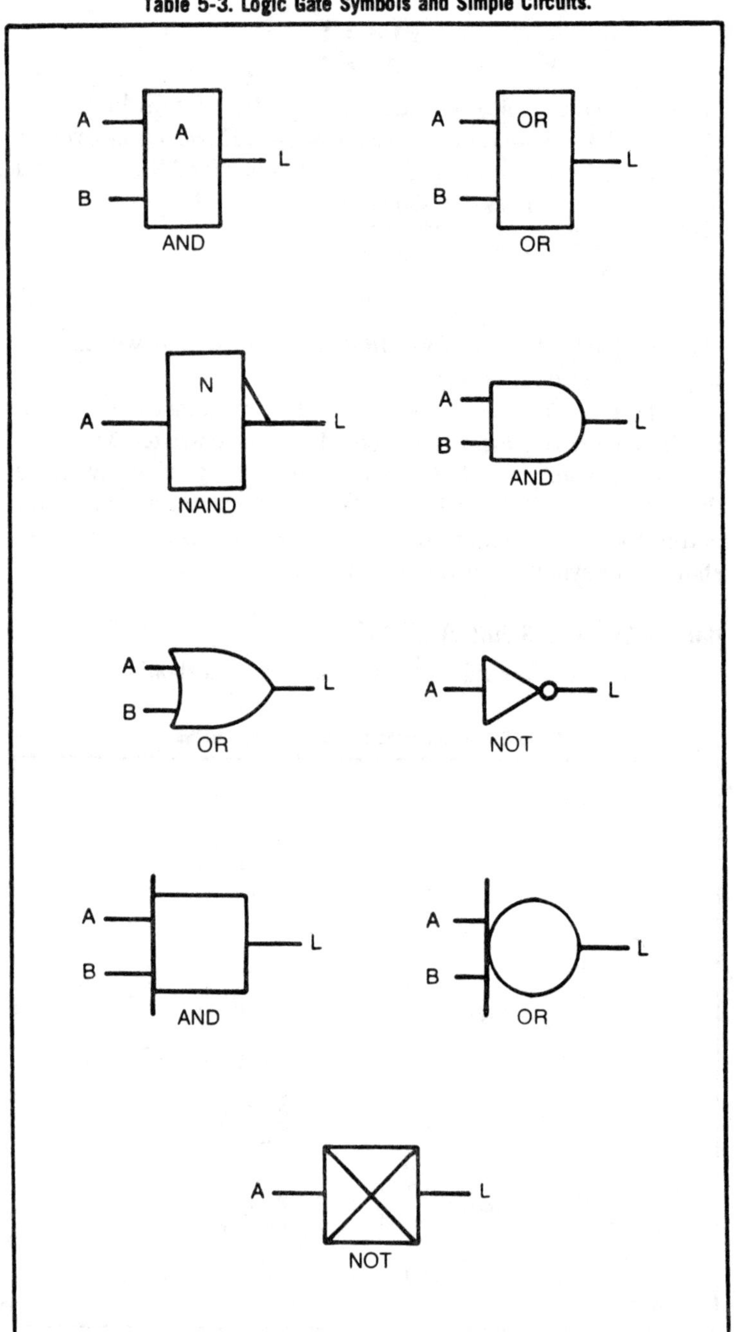

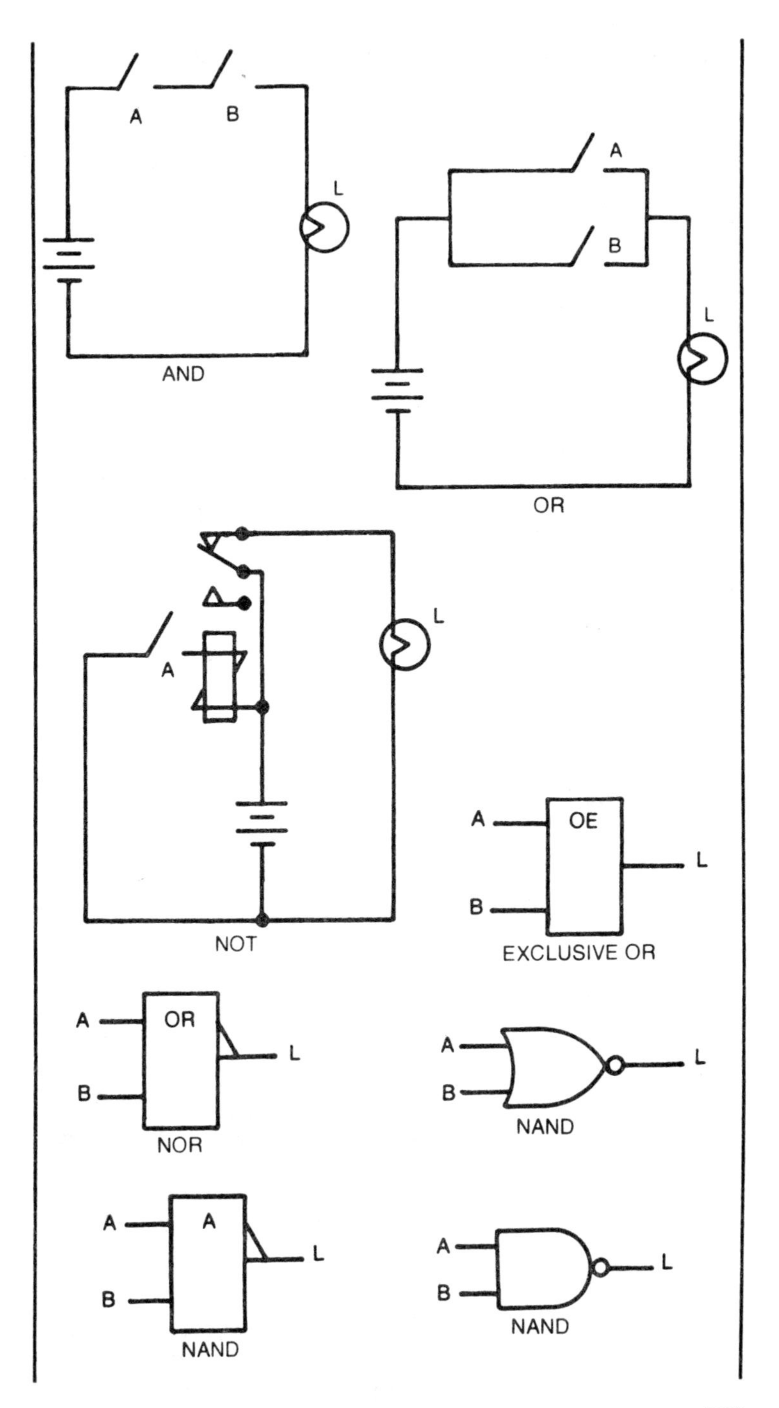
A
B
L
AND
A
B
L
OR
L
A
NOT
A
OE
L
B
EXCLUSIVE OR
A
OR
L
B
NOR
A
L
B
NAND
A
A
L
B
NAND
A
L
B
NAND

Table 5-3. Logic Gate Symbols and Simple Circuits. (Continued from page 109.)

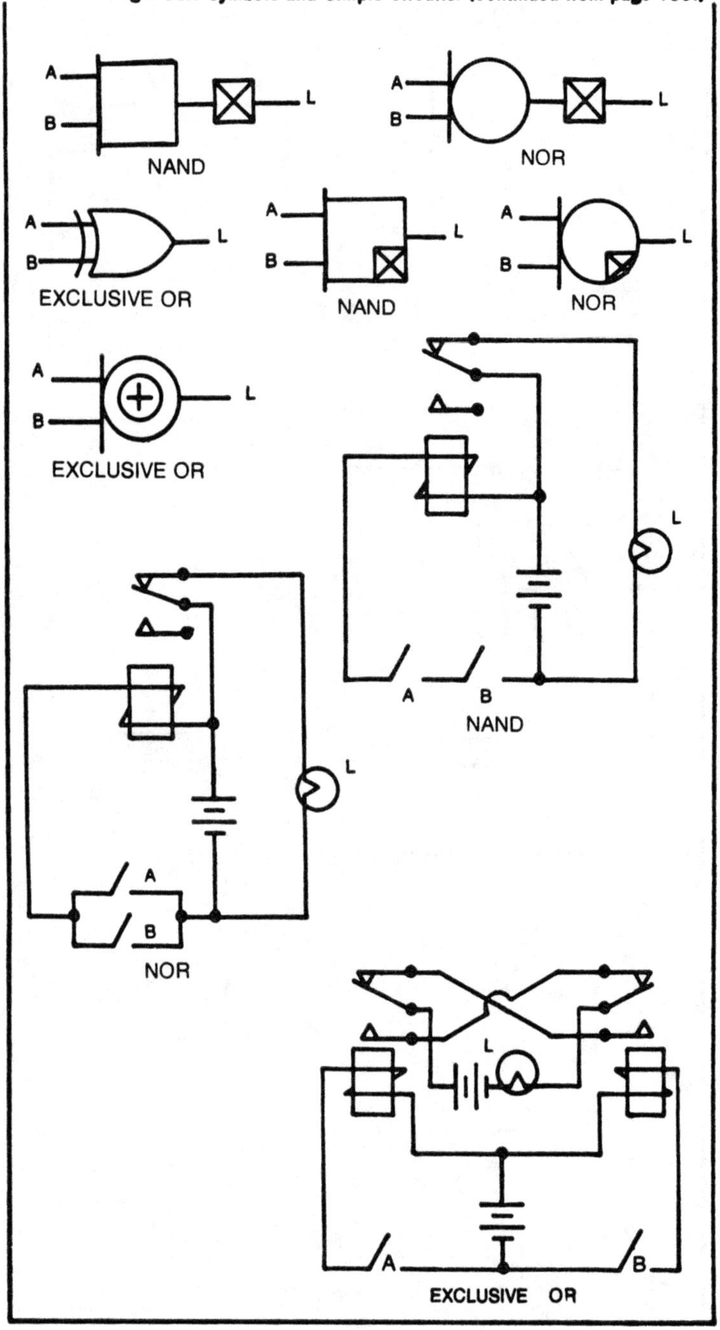

A B L
NAND

A B L
NOR

A B L
EXCLUSIVE OR

A B L
NAND

A B L
NOR

A B L
EXCLUSIVE OR

L
A B
NAND

L
A B
NOR

L
A B
EXCLUSIVE OR

with relation to Boolean algebra. Any time there is an even number of overbars they can be eliminated without changing the meaning of the Boolean expression. For example,

$$\overline{\overline{A}} = A \text{ and } \overline{\overline{\overline{\overline{A}}}} = A$$

If there are an odd number of overbars you can eliminate pairs of overbars without changing the value of the Boolean expression. For example,

$$\overline{\overline{\overline{A}}} = \overline{A} \text{ and } \overline{\overline{\overline{\overline{\overline{A}}}}} = \overline{A}$$

It is useful to know these relationships when you are simplifying Boolean expressions.

There are two Boolean equivalents that you should know. They are referred to as *DeMorgan's theorem*. They are listed here for your convenience. The terms on the left side of the equation are identical to those on the right side of the equation, so these two equalities might be referred to as identities.

$$\overline{A + B} = \overline{A} \times \overline{B}$$

and

$$\overline{AB} = \overline{A} + \overline{B}$$

Technicians use the mnemonic device *break the bar, change the sign* for converting from the expression on the left side of the equation to the expression on the right side of the equation. DeMorgan's rule is very useful for simplifying Boolean equations and also for understanding the simplification of some basic circuits.

Here is a common mistake made by technicians:

$$\overline{AB} = \overline{A + B}$$

This is not true! (Try to apply the rule of *break the bar, change the sign* to either side of this equation and you will see why this equation is *not true!*)

COUNTING SYSTEMS

There are four methods used for counting in digital circuits. You should know all of them and be able to convert between them if necessary. Table 5-4 shows the four methods of counting from 0 to 16.

Note that the decimal 15 is equal to octal 17, and decimal 15 is equal to hexadecimal F. Four ones in the binary are used to represent decimal 15.

The subject is too extensive to go into in a short review, but if you cannot convert between these basic numbering systems, you should take time to learn how to do so before you take the CET Test.

Typical CET Test Question

Convert the following:

$$14_{10} = \underline{\qquad}_{16}$$

The correct answer is

(1) F.
(2) 14_{16}.
(3) E_{16}.

Answer: Choice (3) is correct. The subscript tells the *base* (usually called *modulo*) of the numbering systems. So, E_{16} means the number is in hexadecimal.

As a general rule, most of the conversion problems in the CET Test involve numbers that are less than 32. If you can't do it any other way, learn to write the numbers from 0 to 32 in each basic system.

If you study the columns in Table 5-4 you will very quickly learn that there is a pattern to the way the numbers are increased. For example, when all of the numbers in a certain system have been used, a new column is started. That's why a new column is started for the number 10 in the decimal numbering system. However, in the octal system, where there are only 8 symbols, a new column has to be started after the digit 7 (the eight symbol).

Table 5-4. Number Systems.

Decimal	Binary	Octal	Hexadecimal
0	0000	0	0
1	0001	1	1
2	0010	2	2
3	0011	3	3
4	0100	4	4
5	0101	5	5
6	0110	6	6
7	0111	7	7
8	1000	10	8
9	1001	11	9
10	1010	12	A
11	1010	13	B
12	1100	14	C
13	1101	15	D
14	1110	16	E
15	1111	17	F
16	10000	20	10

It is very easy to learn to write the numbers between 0 to 32 in all of these systems. If you can count that far, you can count as far as you need to beyond 32.

SOME BASIC FACTS ON HARDWARE

It is presumed that you understand how to count the pins on an integrated circuit regardless of whether it is a flat pack, a dip package or a TO-5 case. Always remember the pins are counted *counterclockwise* when the devices are viewed from above and *clockwise* when it is viewed from below. This same rule applies to counting pins in tube sockets, relay sockets, and other electronic components.

To count the pins, you always start at the identifying mark such as a dot or a tab or simply a space between the pins. That tells you where to start the count. Again, remember that it is important that if you are looking at the *top* of the component, you are counting counterclockwise.

Typical CET Test Question

The arrow in Fig. 5-1 is pointing to pin number

Fig. 5-1. What is the pin number indicated by the arrow?

(1) 01
(2) 14

Answer: Choice (2) is correct. You are looking at the top of the IC package, so the count is counterclockwise. Pin 1 is nearest to the side with the identifying notch.

Figure 5-2 shows two important three-state devices which are so important to your knowledge of basic gates. *The purpose of these devices is to provide isolation between the input and output when the three-state input is at logic 0.* When the three-state input goes to a logic 1, then the input and output are coupled as a normal gate.

The example shown in Fig. 5-2A is called a *three-state buffer.* It does not *change* the signal between the input and output. (In some cases, buffers do increase the voltage or current as necessary for a particular design, but in most cases the buffer is simply a straight-through device.)

The gate shown in Fig. 5-2B is a *three-state inverter*. As with the three-state buffer, there is no output from this gate when the three-state terminal is at logic 0. When the three-state terminal is changed to a logic 1, then whatever level is on the input terminal will be inverted. The inverted level will appear on the output terminal. Three-state devices are very important on bus lines in systems.

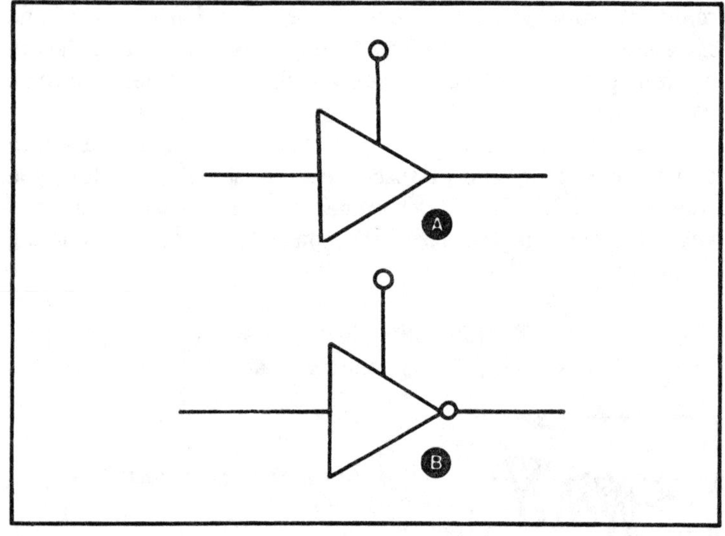

Fig. 5-2. Two important three-state buffers. They are also called tri-state buffers.

Typical CET Test Question

What is the logic level on pin 12 of IC number 2 in Fig. 5-3?

(1) Logic 1
(2) Logic 0
(3) Neither choice is correct.

Answer

The correct choice is (3). Actually, the output of the three-state buffer is an open circuit. That is *not* the same as saying logic 0! In practice, the pin may float to a logic 1 or 0, but from the information in the drawing, you could not say choice (1) or choice (2) is correct.

Very often signals will appear on a bus line that is common to two or three different integrated circuit devices. By using three-state devices at the outputs, they will not be affected by any signal on the line when their three-state terminal is in a logic 0 position. A

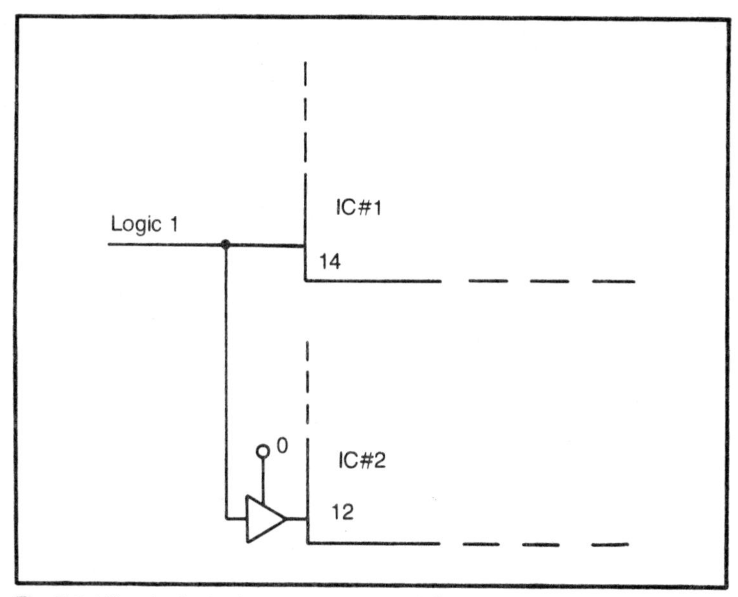

Fig. 5-3. What is the logic level on pin 12 of IC2?

bus is simply a combination of conducting wires or printed circuit conductors that transfer logic signals from one point to another.

FLIP-FLOPS

Flip-flops are very important in logic systems. They are used in many applications. One important application is as memory components or cells. The characteristic of a flip-flop that makes it useful in that application is that it remains in whatever position it is placed until an *imbalance in its input* changes its output condition.

This is best understood by carefully studying the flip-flop shown in Fig. 5-4. This is a simple RS type that can be made with two NAND or two NOR gates. The letter S stands for SET; and, the letter R stands for RESET. When the Q is at a logic 1 and the Q is at a logic 0 (these are the two outputs of the flip-flop) it is considered to be in a HIGH or SET condition. In this particular case, the flip-flop is stable when there are two logic 1's delivered to the S and R terminals.

In 'B' the flip-flop is switched to a LOW or RESET condition by placing logic 0 on the R. When the R is returned to logic 1, the flip-flop remains in the LOW condition.

Delivering another 0 to the R terminal as shown has no effect on the condition of the flip-flop. The only way it can be changed is to

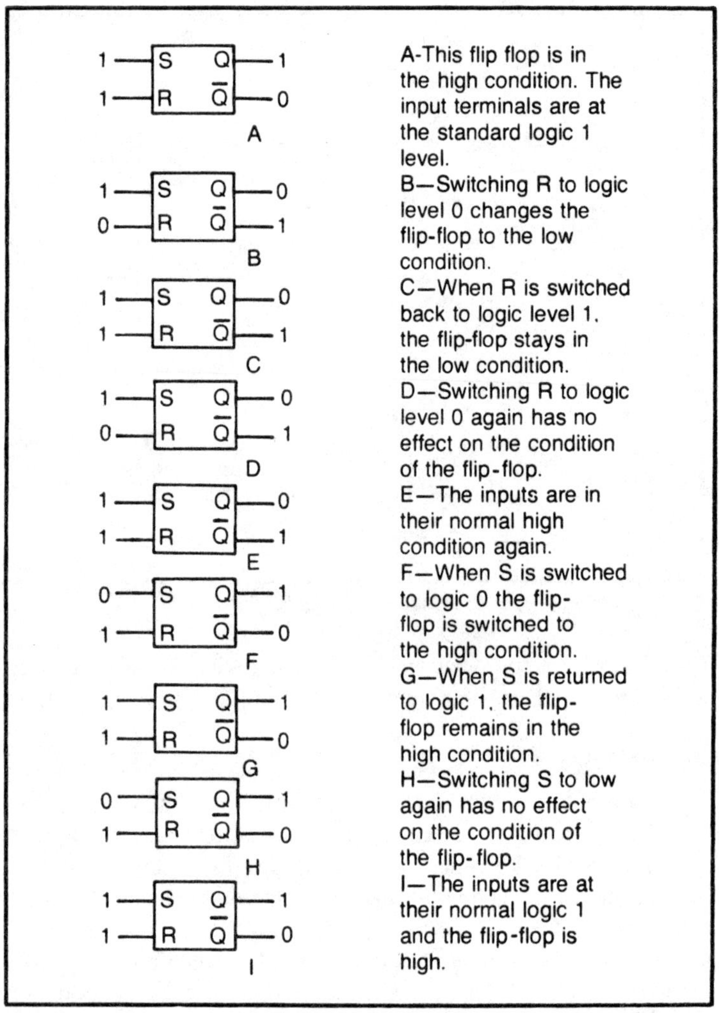

Fig. 5-4. This sequence of events explains the operation of an RS flip-flop.

deliver an unbalancing signal (logic 0 in this case) to the SET input. When the logic 0 is delivered to the SET terminal, it switches the flip-flop to a HIGH condition. In 'G' the flip flop is again returned to its stable input with two logic 1's. Note that in 'H' when a 0 is again delivered to S, the condition of the flip-flop cannot be changed. It could only be changed by a signal now to the R terminal. In 'I' of the illustration, the flip-flop is returned to a highly stable condition with two logic 1 inputs. An important feature of this flip-flop is that it must not have two logic 0's delivered to it at the same time.

In another version of the flip-flop, which is made with NOR's instead of NAND's, it requires two logic 0's in to the S and R terminals to be in a stable state. In that case, the flip-flop's condition is changed by delivering a 1 to the appropriate S or R terminal. *If the flip-flop is stable with the two logic 0 inputs, then it must never be presented with two logic 1's on the S and R terminals.*

An RS flip-flop can be used for storing bits in memory. The flip-flop is placed in a 0 or 1 condition (HIGH or LOW) depending upon which digit (0 or 1) is to be memorized. *It will hold that position for any desired length of time.* However, if the power supply is removed, then the stored number is lost. This type of memory component is referred to as being *volatile*. A non-volatile memory will retain the stored digit even though the power supply is turned off.

Another kind of flip-flop—called the J-K—is used extensively in counting circuits. This flip-flop can be wired so that it will *toggle* (or divide by two) whenever a pulse is delivered to the clock input terminal. If the J-K flip-flop is made with TTL logic, it will normally toggle on the *trailing* edge of an input pulse to the clock input. If it is a CMOS flip-flop, it normally toggles on the *leading* edge of the clock pulse. This is very important because it means *you cannot mix TTL and CMOS flip-flops in a counting system.*

The truth table for a J-K flip-flop is shown in Fig. 5-5. An important thing to understand about this device is that the SET and

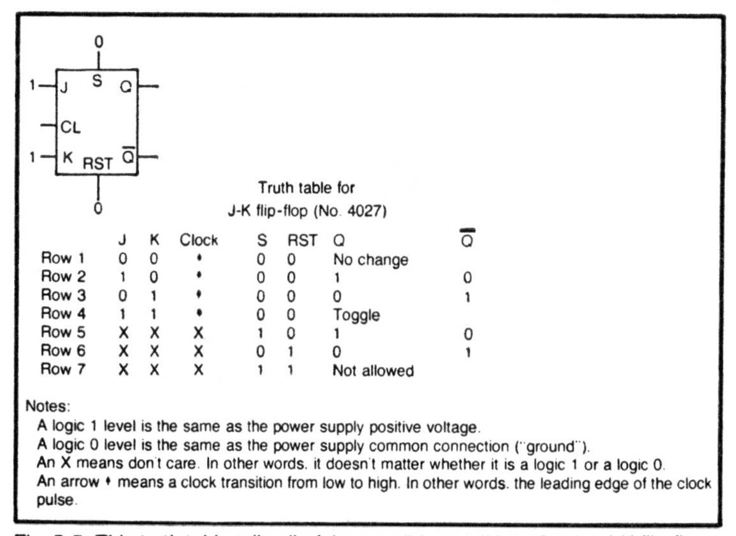

Truth table for J-K flip-flop (No. 4027)

	J	K	Clock	S	RST	Q	$\overline{Q}$
Row 1	0	0	↑	0	0	No change	
Row 2	1	0	↑	0	0	1	0
Row 3	0	1	↑	0	0	0	1
Row 4	1	1	↑	0	0	Toggle	
Row 5	X	X	X	1	0	1	0
Row 6	X	X	X	0	1	0	1
Row 7	X	X	X	1	1	Not allowed	

Notes:

A logic 1 level is the same as the power supply positive voltage.
A logic 0 level is the same as the power supply common connection ("ground").
An X means don't care. In other words, it doesn't matter whether it is a logic 1 or a logic 0.
An arrow ↑ means a clock transition from low to high. In other words, the leading edge of the clock pulse.

Fig. 5-5. This truth table tells all of the possible conditions for the J-K flip-flop.

RESET terminals control the flip-flop regardless of the logic levels of J or K or the clock.

Let's look at each row of the flip-flop truth table and make sure the meaning is clear. In Row 1, having J and K and SET and RESET all at logic 1 means that no change can take place even though there is a clock transition from LOW to HIGH. In row 2, the flip-flop will be HIGH when J is 1, K is 0, and the clock transition is on the leading edge. Note that S and R terminals are at logic 0 which is their stable condition in this flip-flop. In Row 3, the flip-flop is switched LOW by making J a logic 0 and K a logic 1. Again, this occurs on the positive clock transition. In Row 4, J and K are both made equal to logic 1. If S and R are both at 0 as shown in the truth table, then the flip-flop will toggle, or divide by two. In Rows 5 and 6, you can see that the J, K, and clock are in a don't-care condition because the inputs to the S and R terminals control the flip-flop. Row 7 shows that the SET and RESET terminals must never be placed at logic 1 at the same time.

The data flip-flop shown in Fig. 5-6 is much more convenient to use as a memory component. Assuming the SET and RESET terminals are logic 0, then a logic 1 at the D input will produce a HIGH condition for the flip-flop. You can see this in the first row of the D table flip-flop. However, simply placing a logic 1 to the D input is not sufficient to cause the flip-flop to go HIGH. It is necessary at the same time for the clock terminal to go from a 0 to a logic 1. That is the reason why the arrow is pointing UP under the clock count. It simply means that to switch the flip-flop to a HIGH condition, the D must be in a logic 1 and the clock terminal must go from 0 to logic 1.

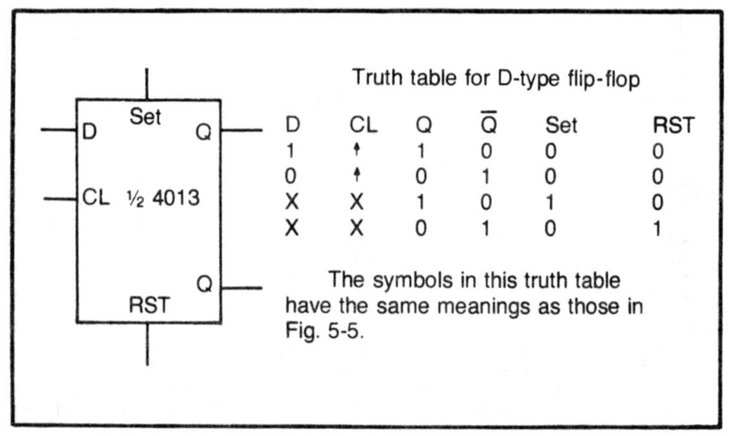

Truth table for D-type flip-flop

D	CL	Q	$\overline{Q}$	Set	RST
1	↑	1	0	0	0
0	↑	0	1	0	0
X	X	1	0	1	0
X	X	0	1	0	1

The symbols in this truth table have the same meanings as those in Fig. 5-5.

Fig. 5-6. This is the symbol used for a data flip-flop. The truth table for this flip-flop is also illustrated.

It is important to note that this flip-flop, like most of them, changes its condition on either the leading edge or the trailing edge. The truth table for the flip-flop will tell you which edge (leading or trailing) changes the flip-flop.

To switch the D-type flip-flop LOW, a 0 must be put to the D terminal, and at the same time, the clock must undergo a transition from 0 to 1 again. If the SET and RESET are *not* at logic 0, then the D terminal has no control over it.

In the third row of the D-type truth table, you can see that when the SET is at logic 1 and the RESET is at logic 0, the flip-flop is HIGH regardless of the condition of the D and clock inputs.

In the fourth row, you can see that if SET is 0 and RESET is 1, the flip-flop is LOW (0 out from Q and 1 out from $\overline{Q}$) regardless of whether the D terminal and the clock terminals are at 1 or 0. The D-type flip-flop, like the J-K type, can be toggled. So, it is sometimes used in counting circuits. As mentioned before, the D-type flip-flop is much more convenient to use as a memory element than the RS flip-flop previously described.

MEMORIES AND MICROPROCESSORS

The use of dedicated microprocessors in television receivers has made a significant change in the types of services being made available to the viewer. A very few examples are given here:

- The time of day can be displayed on the screen.
- The channel number can be displayed on the screen for a brief period when the channel selection is first made.
- The set can be programmed to automatically turn ON to the correct channel for selections made a week ahead of time.
- Push-button selections for the channels have been made available.

How does the microprocessor accomplish all of these things? In the technical literature you see all types of strange characteristics assigned to the microprocessor. It is variously called ***the brain,*** or ***the heart,*** of a digital system. In reality, a microprocessor is so dumb that it is an absolute insult to compare it to a human brain or even the brain of a mouse. And, if anything should be called the heart of a system, it would be the *clock* which all microprocessor and synchronous digital systems have. The clock is responsible for pulsing the data along the data buses, and for producing the continuing sequences of events that occur when the microprocessor is working.

To understand what the microprocessor really does, it is a

good idea to go back and find out for what purpose it was originally designed. Originally, the microprocessor was designed to *implement memory*. Consider the plight of manufacturers of integrated circuit memory. From a relatively crude beginning of only a few memory locations, it very soon became possible to put 64,000 bits of memory on a single chip. Then, the real problems started. How do you get bits of information into and out of the memory without 64,000 wires?

The limit to the number of memories that could be sold was definitely based on the wiring complexity that the user (or designer) was willing to put up with. What was needed was a programmable device which would enable the user to get into and out of memory in some simple way. *That was the reason the microprocessor was first designed!*

It is not uncommon to find literature today that refers to the microprocessor as a computer. I suppose a purist would say that's an insult to the people that make computers. The reason people ascribe this feature to microprocessors is that one of the basic parts of a microprocessor is the *arithmetic logic unit* (ALU). This was put in to the microprocessor so that data could be retrieved from the memory and operated upon before it was presented to the outside world.

The ALU performs simple addition and subtraction and other basic arithmetic operations. It also performs the basic logic operations related to the seven basic gates that were discussed at the beginning of this chapter.

All of the microprocessors in use today can be divided into two basic categories: *dedicated* and *undedicated*. A dedicated microprocessor is on that is programmed in the factory and can be used only for some particular application. The microprocessor used in television tuners, for example, are dedicated microprocessors. Undedicated microprocessors, like the 8080A, 6800, and 68000, are very versatile compared to the dedicated units in television tuners. These microprocessors, the undedicated ones, can be programmed to perform in almost limitless numbers of tasks.

A second category for microprocessors is its *number of bits*. This is actually the number of bits of information that can be sent on the data bus at the same time. All microprocessor systems have at least three buses—the *data* bus, the *address* bus and the *control* bus. The data bus carries the information to the memory and away from the memory. It also carries the information outside of the microprocessor and brings outside information in. This is only true, of course, in microprocessors that are not dedicated.

The number of bits of data on the data bus is a direct determining factor on how many steps an operation must go through in order for a task to be completed. Today, the 16-bit microprocessor is often favored over the 8-bit microprocessor because it can perform tasks much more quickly. The reason this is true is because some of the bits on the data line can be used to identify the operation. If those bits aren't available, then a separate step must be used to identify the operation required. The tradeoff is in the fact that the greater the number of bits on the data bus, the more complex the programming operation, and the more expensive the microprocessor system.

In the television tuners, and in other appliance applications, 4-bit microprocessors are popular. The reason 4-bit microprocessors are used is simple. There just isn't any need for any more bits. Well, at least that was the original contention. Because of the limited number of pins on a 4-bit microprocessor, there is some difficulty in getting all of the outside data into the chip itself.

The way the designers get around this problem is to use what is known as a *multiplexer.* A mulitplexer is simply a logic circuit that permits a lot of inputs to be reduced to a single output. A *demultiplexer,* on the other hand, takes a single input and divides it up among a lot of outputs. These multiplexers and demultiplexers make it possible to use operational microprocessors that have less pins available than the higher bit models. If you look at a microprocessor system as it is used in a computer, you will see that there are buses from the microprocessor to the various memories. They are shown in Fig. 5-7.

The *random access memory* (RAM) which is a temporary storage device, is volatile. It is specifically designed to have the information stored in that memory changed frequently. The *read only memory* (ROM) has the information stored during manufacturing. That information cannot be changed, and it is non-volatile.

As microprocessor systems developed, there emerged a need for read only memories that were non-volatile but could be programmed in the field. In other words, the user did not want the memory programmed for eternity. That led to the development of the PROM, or *programmable read only memory.* With this type, the original stored information could be erased with ultraviolet light, and the memory could be reprogrammed in the field.

Using all of that equipment to reprogram the memory was inconvenient so *Electrically Erasable Programmable Read Only*

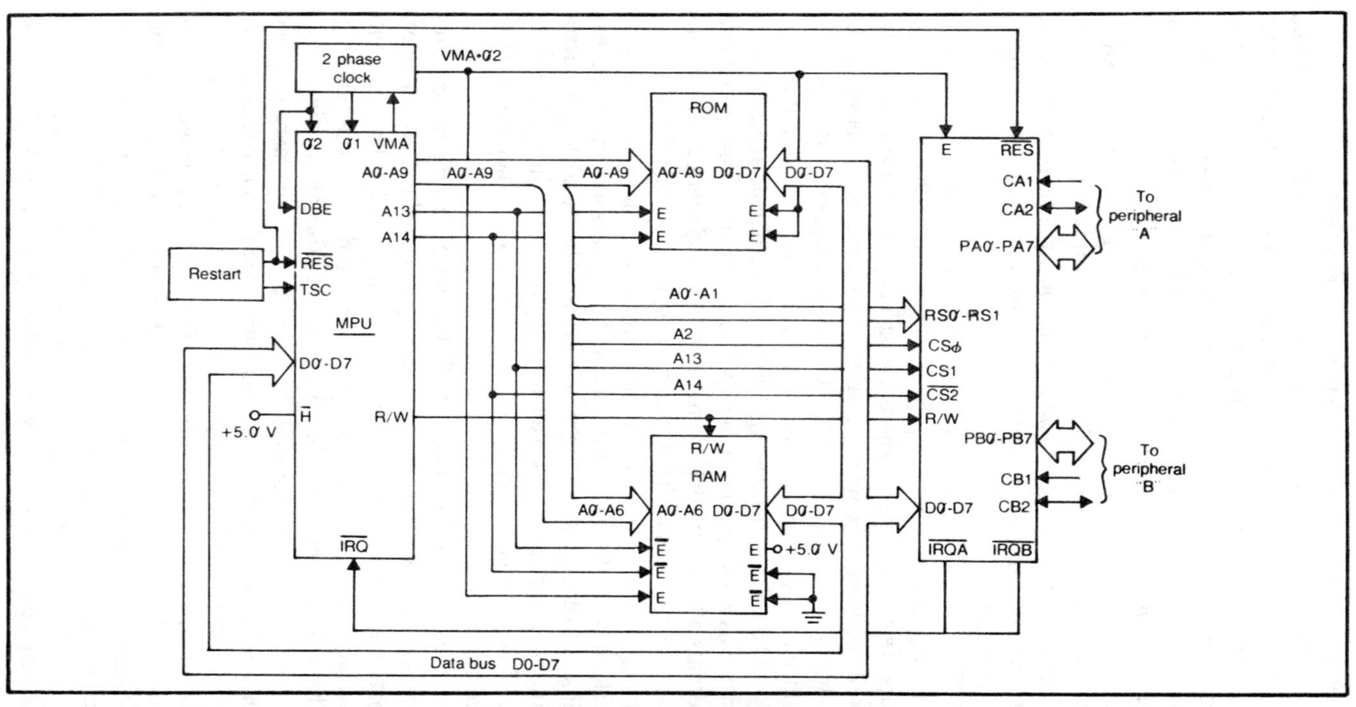

Fig. 5-7. Buses are used to interconnect the microprocessor with memories. This system would be located within a computer.

Memories (EEPROM's) were developed. These are very popular in television systems. For example, when a customer moves into a new area with his television set, he wants to program his TV receiver to receive the channels available. Once he's programmed it, he does not want to have to do that again (until he moves again). So, the read only memory used for memorizing the various channels was made to be electrically erasable.

AN INTRODUCTION TO MICROPROCESSOR TERMINOLOGY

A good way to get familiar with microprocessor terminology, is to review the pinout of a typical one. Figure 5-8 shows the pin designations for a microprocessor in the 6800 (Motorola) family. *Pin 8* is marked VCC. It is a dc power supply connection for the microprocessor. As with many other electronic components, the microprocessor will not operate without a positive and negative dc supply voltage.

Now, while we are at the supply connections, look at pins 1 and 21 and you will see that they are marked VSS. They are the common or "ground" connections for the microprocessor. So, the power supply must be connected between pins 1 and 21, and pin 8. Why would a manufacturer put VSS on both sides of the integrated circuit? The reason is simple. It makes it easier to design printed circuit boards. A big problem in printed circuit boards is to design them so that conductors aren't crossing over one another. It might be impossible to get the VSS on one side of the microprocessor, but possible on the other side. So, the manufacturer has simplified the mounting of this device on printed circuit boards.

Pin 2 is marked with a $\overline{\text{HALT}}$. Any time you have a line over the top of a microprocessor or digital symbol, it means NOT or NO. Therefore, pin 2 is a NOT HALT input pin. What that means is that there must be a logic 1 into this pin to prevent the microprocessor from stopping what it is doing.

The next pin shows a phase 1 (Ø1) terminal. If you look over to pin 37, you will see the phase 2 (Ø2) terminal. These two terminals provide the clock input to the microprocessor.

Everything in the microprocessor runs with a clock signal, which is nothing more than a square wave. So, you would expect to see a square wave on pins 37 and 3 when you scope those points. If you looked at them both at the same time with a dual-trace scope, you would find that they are approximately 180° out of phase. Not all microprocessors use two-phase clocks, but they will all have at least one clock.

Pin 4 is an $\overline{\text{IRQ}}$. This means that it is a *Not Interrupt Request.* An interrupt request comes from outside the microprocessor. It is simply a way of telling the microprocessor that there is information available that should be used and as soon as the microprocessor can get to it, would it please service the request.

Pin 5 is marked VMA. It is a *Valid Memory Address.* You have to understand that memories cannot distinguish between data pulses and an undesired pulse. The VMA pin is used by the microprocessor to tell the memory this is really a signal, it is not fooling.

Pin 6 is an $\overline{\text{NMI}}$ identification. The $\overline{\text{NMI}}$ identification means *Non-Maskable Interrupt.* This is an interrupt that has to be taken seriously by the microprocessor. You can't shut this interrupt out. It is not going to go away. This is an important interrupt signal to the microprocessor. By contrast, a plain interrupt tells the microprocessor that it has information to be serviced whenever it is convenient for the microprocessor.

Pin 7 is marked *Bus Available* (BA). This pin is like a streetcar conductor or traffic cop. You wouldn't want to try to send signals into the microprocessor and out of the microprocessor both at the same time. The BA pin will prevent that.

The VCC at pin 8 is for the positive 5 volts that is used for operating the microprocessor.

Now, look at pins 9 through 20 and 22 through 25. There are a total of 16 pins marked with an A identification number. The letter A stands for *Address.* Keep in mind the fact that you can only use logic 1's and 0's for information; and, you have 16 lines to carry the information. It follows mathematically that you have 2^{16} or 65, 536 different codes or addresses, that you can identify on the 16 bus lines. Another way of saying this is that this microprocessor can direct signals to 65,536 different address locations. The address is the place where information is stored in memory.

In microprocessor language, the number 65,536 is called 64 k (64,000). Usually a person studying this for the first time wants to know why they call it 64 k instead of 65,536. I don't really know the answer, but I think it goes back to the time when these types of memories were first being used. The manufacturer of the memories only guaranteed so many memories of the total. In other words, if you were buying a memory that was supposed to be able to store information at 16 addresses, you might only get 15. So, they used a lower number that they were willing to guarantee.

Look at pins 26 through 32. Here are eight pins marked with a D identification. The D stands for *Data.* The microprocessor is

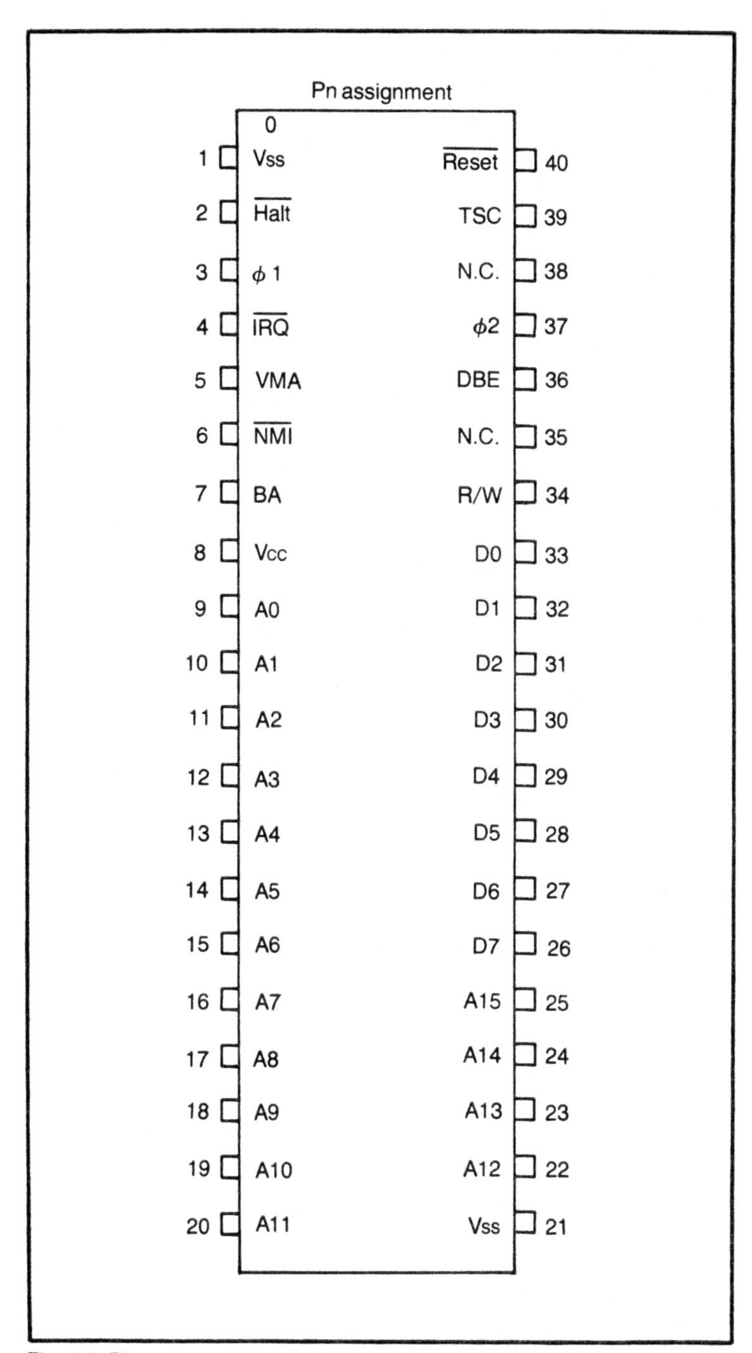

Fig. 5-8. Pinout for a 6802 microprocessor.

involved with two different kinds of signals. One, called the *address,* tells where the information is going to or coming from. The other, called the *data,* tells that the actual information is in the form of a code. Let me give you an example. Suppose you wanted to store the number 10 in location 0015. The first thing you would need would be the location where you want to store it. That is the 0015 address. The second thing you would need is the number, or data. In this case, the data to be stored is 10.

Pin 34 is a *Read/Write* (R/W) pin. This pin is used for operating certain types of memory. It tells the memory whether the μP is putting information (data) *into* memory or taking information (data) *out of* memory. In this sense, the memory is very much like your brain. If you want to put data in, that is one thing. If you want to take data out, that's another. Most of us can't do both things at the same time, and neither can the microprocessor. The R/W pin (34) prevents this from happening.

Pin 35 (NC) has no connection. Pin 38 is also NC.

Pin 36 is marked DBE which stands for *Data Bus Enable.* That pin saves a lot of burned-up boards. It simply prevents the information from coming in and going out at the same time. It does that by sending enabling and disabling signals to 3.

Pin 39 is identified as TSC. That stands for *Three-State Control.* A three-state control enables the μP to make its data pins open circuited so that information can't be pushed in or out. So, the μP is protected against signals being delivered to the data and address pins (and others) when the μP isn't ready.

Suppose, as an example, that data input at pin 33 is at logic 1; and, at the same time, that pin is at logic 0 due to an internal μP action. Now you have 5 volts and 0 volts both connected to the same pin. This is technically a short circuit. (Something has got to give). With the three-state control, it makes the pins open circuited and this situation is avoided.

The last pin (No. 40) is marked $\overline{\text{RESET}}$. When you first turn the μP ON, there are various internal sections of the μP which will be at 1's and 0's in a sort of random arrangement. It is necessary to get all of these signals clear (or RESET) before using the μP. This is what a signal on the RESET pin does.

Typical CET Test Question

A microprocessor is going to store 11001010 in the memory at 0000000011100011. Which of the following is correct?

(1) The address is 11100011* and the data is 11001010.

(2) The address is 11001010 and the data is 11100011*.

Answer: Choice (1) is correct. Note that the data is 8 bits long and the address is 16 bits long. This is typical for an 8-bit microprocessor.

OTHER IMPORTANT INTEGRATED CIRCUIT LOGIC SYSTEMS AND DEVICES

Inside of the microprocessor there is a necessity for very temporary storage of data while that data is waiting to be operated upon by the ALU. This data is stored in a section called the *register*.

Registers are also used in logic systems outside of the microprocessor. All registers can be divided into two basic classifications. Those that have *parallel data* in which all of the data goes in or comes out at the same time. The other type has *serial data* in which the data comes in and goes out on a single line. Combinations of serial and parallel are possible as shown in Fig. 5-9.

*It is customary to disregard the first eight 0's of the address.

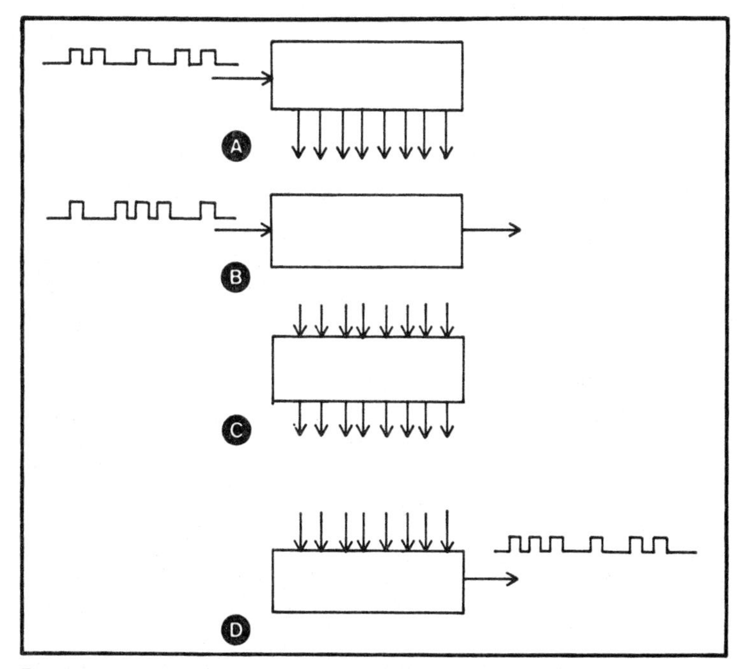

Fig. 5-9. Various configurations are possible with registers: (A) Serial In-Parallel Out (B) Serial In-Serial Out (C) Parallel In-Parallel Out (D) Parallel In-Serial Out.

A moment's reflection will show you that a serial data system is necessary if you are going to send an 8-digit number over a telephone line. Since there is only one single line, it is necessary to take the 8 digits and break them down into a code that allows one digit to be sent at a time. So, serial registers are used extensively for getting data in and out on telephone lines. Parallel registers are used extensively in microprocessors and other systems where the data can be delivered 8 bits at a time. (This, of course, assumes that it is an 8-bit system.) In a 16-bit system, the register would have to have 16 parallel bits.

If you were able to take apart the dedicated microprocessor from a tuner, you would find that it is not just simply a microprocessor. In most of these integrated circuits, there is also a RAM and the ROM. To get all of that into a package, a very large number of pins is necessary on the integrated circuit chip. It is not uncommon to find 60 pins!

Something else you might find in the microprocessor IC would be a phase-locked loop. This circuitry will be discussed in the next chapter on linear circuits, but for now it is important to understand that you might find one inside the microprocessor. They are used for tuning the receiver to a station and holding that station while the viewer is watching it.

There's a very important thing to remember about integrated circuits with so many pins. Technicians report that they are having trouble getting them in and out of the circuit board because *they have a tendency to break in half when you pry on one end of them!* You're not even guaranteed against breakage by prying at both ends at the same time, but that is the least risky way to remove microprocessors from printed circuit boards. Of course, before you remove them, they usually have to be desoldered because it is not a common practice to use integrated circuit sockets.

In one of the CET Tests, the term *bit slice* was mentioned and a large number of letters were received asking just what is a bit slice. Well, if you look at the block diagram of an integrated circuit microprocessor, you will find that it is made up of various individual sections like the *ALU,* the *registers,* the *program counter,* etc. It should be obvious that these various sections could be located in individual integrated circuit packages rather than in a single package like the microprocessor. When you do that (locate the various sections of a microprocessor system in separate packages), you have put together what is known as a *bit slice.*

What is the advantage of bit slicing? Simply that it is the fastest microprocessor system available today. The thing that makes a regular microprocessor slow compared to a bit slice, is the fact that the data has to be moved out and in to the various sections of the microprocessor system. A bit-slice design has no unnecessary sections, and only the absolutely necessary data motion is utilized in performing a desired function.

Bit-slice design has a great disadvantage in that it is much more difficult to program. You will *generally not see* bit-slice design in microprocessors used in television tuners. They are more readily found in industrial applications.

ADDITIONAL TERMS YOU SHOULD KNOW FOR THE CET TEST

Synchronous and Asynchronous Circuitry. A synchronous circuit is one that requires the presence of another signal for its operation. The second signal is usually referred to as a clock pulse. By using a clock signal, it assures that circuits are not accidentally triggered into operation by noise pulses.

Glitches. Glitches are undesired and unwanted signals of very short duration (like transients). Glitches are produced in a circuit when the timing between two basic components is not exactly correct. They have a very bad habit of destroying circuit operation.

TTL, CMOS, and ECL Logic Circuitry. The initials *TTL* stand for transistor-transistor logic. This type of logic is very often numbered in the 7400 or 5400 series. The difference between these is in the tightness of the specifications rather than in the pinout or mode of operation. Thus, a 7400 is basically the same circuit as a 5400. As a general rule, the 5400 devices are designed for military operations with tight specifications, while the 7400 devices are used in commercial equipment.

One special branch of the TTL family is the *Schottky* devices which have an extremely fast switching time. As a matter of fact, it is possible to obtain switching times (or propagation delays) of 3 nanoseconds per gate. That's not the fastest. *Emitter-couple logic* gates are the absolute fastest of the integrated circuit families.

CMOS devices are made with combinations of N-channel and P-channel MOSFET's. They are usually numbered in the 4000 series and can operate with voltages between 3 volts and 15 volts. They are not as fast as TTL's, and you will see typical propagation delays of 15 to 35 nanoseconds. Probably the most important advantage of CMOS devices is that they do not require a regulated 5-volt supply. There has been some problem—especially in early CMOS integrated circuits—with static electricity. However, that problem is not serious in newer components, but care should be exercised to avoid unnecessary exposure to static charges. The abbreviation *ECL* stands for *emitter-coupled logic.* This very fast logic family has propagation delays as little as one and a half nanoseconds.

An important characteristic of emitter-coupled logic is that they provide a constant drain on the power supply, and therefore, do not produce switching transients. Another important characteristic is that they are operated with a negative 5 volt power supply. ECL logic gates are normally numbered in the 10,000 series.

Programmable Counters. A very important feature of many logic systems is a counter. Those for low-number counts (or modulos) can be made with toggled flip-flops (remember that either D or J-K flip-flops can be readily toggled).

For higher counts, manufacturers have put into an IC package, *programmable counters.* These are characterized by the fact that they can count up to high numbers, or count down from high numbers. Also, they can be set to count up or count down to a desired value by applying the proper logic levels to the programmable data inputs.

Synchronous and Asynchronous Counters. The terms *synchronous* and *asynchronous* are sometimes applied to counters. For the synchronous counter, all the flip-flops change their state at the same time when a count is being made. This usually produces a rather heavy burden on the power supply. However, since they all change at the same time, the synchronous counter is much faster than the asynchronous counter.

When an asynchronous counter adds a digit or subtracts a digit from the total count, it is necessary for a change to be produced by signals which ripple through each individual flip-flop. Thus, asynchronous counters are sometimes referred to as ripple counters or ripple-through counters.

The fact that not all of the flip-flops change their condition at the same instant means that there is less drain on the power supply. However, it takes longer for a count to be made because the flip-flops do not all change their condition at the same instant of time.

Encoders and Decoders. Encoders and decoders are used to translate numbers from one digital system into another. An example of the use of an encoder would be to change the pressure on a typewriter key into the ASCII code which can be recognized in a microprocessor system. A decoder example would be an integrated circuit that converts a binary count into signals for operating a seven-segment readout.

Static Versus Dynamic Memory. A *static memory* is one which holds a binary number over a period of time and does not require refresh signals. Flip-flops and latches are used to produce static memories.

A dynamic memory is one which requires a continuous recirculating signal to maintain the memory count. This is called the refresh signal. Dynamic memories are sometimes made by charging capacitors. The charged capacitor indicating that a binary 1 is present and the uncharged capacitor represents a binary 0. Once a capacitor is charged, it has to be continually recharged or refreshed to maintain the memory count.

In comparing dynamic and static memories, you should understand that dynamic memories are much faster for access and it is possible to put a very large number of memory cells on a single integrated circuit. The disadvantage is that the refreshed circuit requires a constant recharging energy.

Charge-Coupled Devices (CCD). Charge coupling occurs when static charges are stored under the surface of certain types of material. These charges can represent 1's and 0's depending upon whether or not they are present. The characteristic of charge coupling that is most important is the high density (or number of memory cells per chip) and the very rapid speed with which the information stored can be entered and retrieved.

A/D and D/A Converters. Analog-to-digital and digital-to-analog converters are necessary because much of the electrical information available in the outside world is not suitable for operating microprocessor and digital systems. Therefore, the data has to be converted from an analog (linear) format into a digital format so that it can be utilized in the microprocessor and digital systems. Conversely, if the information is to be sent to the outside world, the digital information may not be suitable for operating linear systems; and, the digital signal must be converted from digital to analog.

PROGRAMMED REVIEW

Start with block number 1. Pick the answer that you feel is correct. If you select choice number 1, go to Block 13. If you select choice number 2, go to Block 15. If you select choice number 3, go to Block 28. Proceed as directed. There is only one correct answer for each question.

BLOCK 1

The truth table in this block is for

(1) a 4-input NAND.
Go to Block 13.

(2) a 4-input NOR.
Go to Block 15.

(3) a 4-input EXCLUSIVE OR.
Go to Block 28.

A	B	C	D	L
0	0	0	0	1
0	0	0	1	0
0	0	1	0	0
0	0	1	1	0
0	1	0	0	0
0	1	0	1	0
0	1	1	0	0
0	1	1	1	0
1	0	0	0	0
1	0	0	1	0
1	0	1	0	0
1	0	1	1	0
1	1	0	0	0
1	1	0	1	0
1	1	1	0	0
1	1	1	1	0

BLOCK 2

Your answer to the question in Block 36 is not correct. Go back and read the question again and select another answer.

BLOCK 3

The correct answer to the question in Block 9 is choice (1). You cannot change the program in this microprocessor.

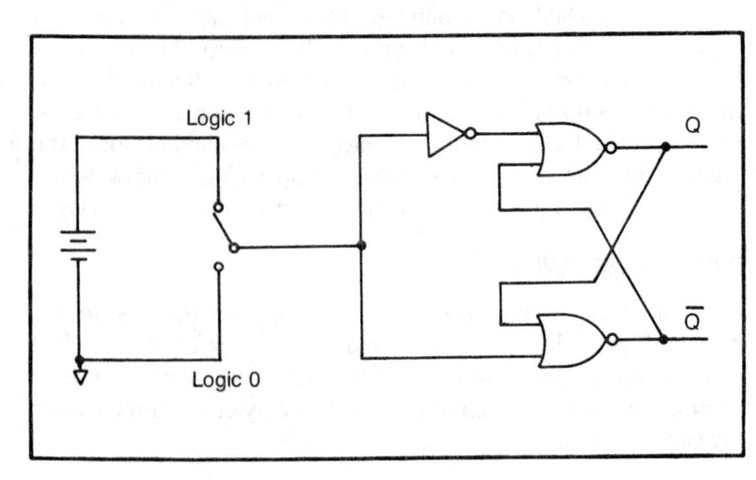

Here is your next question:

What is the condition of the simple D-type flip-flop shown in this block?

(1) HIGH ($Q = 1, \overline{Q} = 0$). Go to Block 39.
(2) LOW ($Q = 0, \overline{Q} = 1$). Go to Block 11.

BLOCK 4

The correct answer to the question in Block 37 is choice (4).

Here is your next question:

The gate in this illustration can be represented by the Boolean equation.

(1) $AB + \overline{AB} = L$. Go to Block 33.
(2) $\overline{A}B + A\overline{B} = L$. Go to Block 6.

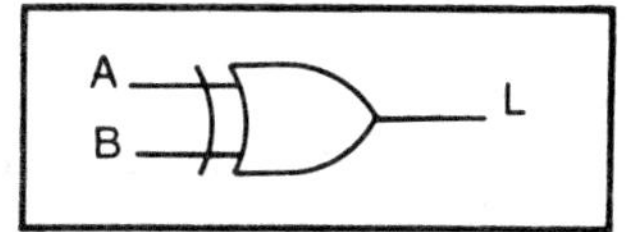

BLOCK 5

Your answer to the question in Block 19 is not correct. Go back and read the question again and select another answer.

BLOCK 6

The correct answer to the question in Block 4 is choice (2). The other selection is the Boolean equation for a logic comparator.

Here is your next question:

The output of the circuit in this block can be written in Boolean equation form as

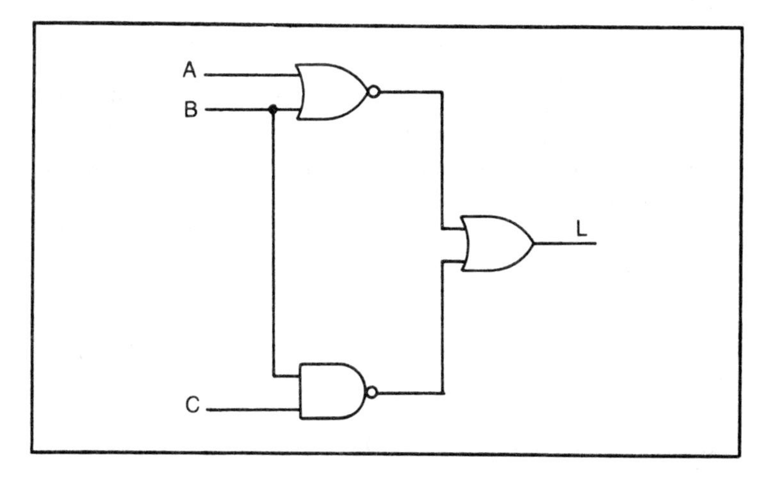

(1) $\overline{BC}$. Go to Block 9.
(2) $\overline{AB} + \overline{C}$. Go to Block 14.
(3) $\overline{AC}$. Go to Block 23.
(4) $\overline{AB}$. Go to Block 18.

BLOCK 7

Your answer to the question in Block 15 is not correct. Go back and read the question again and select another answer.

BLOCK 8

Your answer to the question in Block 10 is not correct. Go back and read the question again and select another answer.

BLOCK 9

The correct answer to the question in Block 6 is choice (1). The simplification shown here starts with the output at L.

$$L = \overline{A + B} + \overline{BC}$$
$$= \overline{A} \times \overline{B} + \overline{B} + \overline{C}$$
$$= \overline{B}\,(\overline{A} + 1) + \overline{C}$$
$$\text{But } (\overline{A} + 1) = 1$$
$$L = \overline{B} + \overline{C} = \overline{BC}$$

Here is your next question:

The microprocessor used in a television tuner is

(1) dedicated. Go to Block 3.
(2) pledged. Go to Block 24.

BLOCK 10

The correct answer to the question in Block 22 is choice (3). At no time do all of the input signals go to a logic 1 at the same instant. The only way to get a logic 0 out of a NAND is to have all of the inputs at a logic 1. So, the inputs at any time are combinations of logic 1 and logic 0 levels, and, the output must be a logic 1 level.

Here is your next question:

With the logic levels shown on the gate in this block, is it possible for the gate to be working properly?

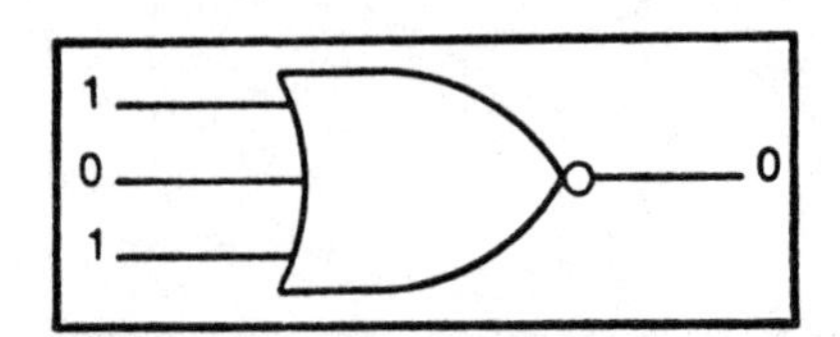

(1) Yes. Go to Block 35.
(2) No. Go to Block 8.

BLOCK 11

Your answer to the question in Block 3 is not correct. Go back and read the question again and select another answer.

BLOCK 12

Your answer to the question in Block 37 is not correct. Go back and read the question again and select another answer.

BLOCK 13

Your answer to the question in Block 1 is not correct. Go back and read the question again and select another answer.

BLOCK 14

Your answer to the question in Block 6 is not correct. Go back and read the question again and select another answer.

BLOCK 15

The correct answer to the question in Block 1 is choice (2). The only way to get a logic 1 out of a NOR gate is to have all inputs at a logic 0 level.

Here is your next question:

Which of the circuits in this block can be described by the Boolean equation $A + B = L$?

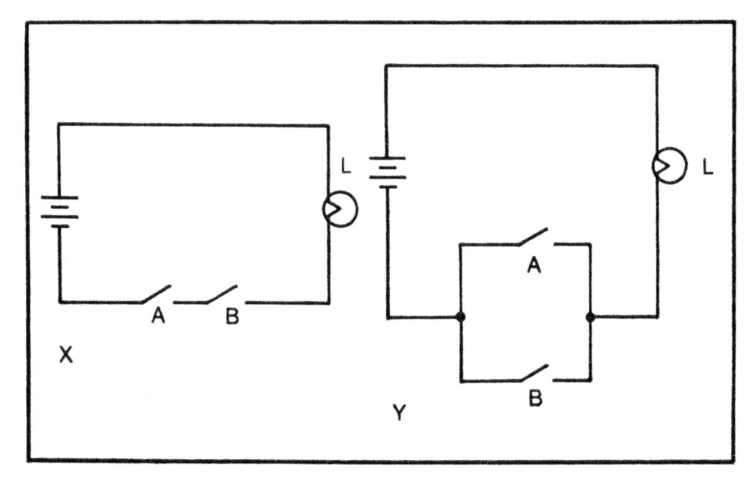

(1) The one marked X. Go to Block 7.
(2) The one marked Y. Go to Block 22.

BLOCK 16

Your answer to the question in Block 39 is not correct. Go back and read the question again and select another answer.

BLOCK 17

Your answer to the question in Block 25 is not correct. Go back and read the question again and select another answer.

BLOCK 18

Your answer to the question in Block 6 is not correct. Go back and read the question again and select another answer.

BLOCK 19

The correct answer to the question in Block 35 is choice (2). The lamp can be operated by switching A or B, but not both.

Here is your next question.

The circuit in this block can be represented by the Boolean equation:

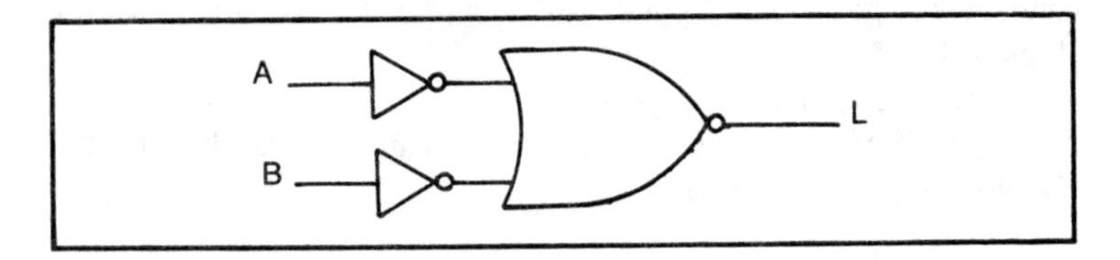

(1) A + B = L. Go to Block 29.
(2) AB = L. Go to Block 37.
(3) Neither choice is correct. Go to Block 5.

BLOCK 20

Your answer to the question in Block 37 is not correct. Go back and read the question again and select another answer.

BLOCK 21

Your answer to the question in Block 22 is not correct. Go back and read the question again and select another answer.

BLOCK 22

The correct answer to the question in Block 15 is choice (2). The equation says that A or B produces an output L. With the circuit shown in Y, either switch will turn on the lamp. In other words, A or B will light the lamp. This is an example of an INCLUSIVE OR circuit because either or both A or B equals L.

Here is your next question:

With the input signals shown, the output at L will be

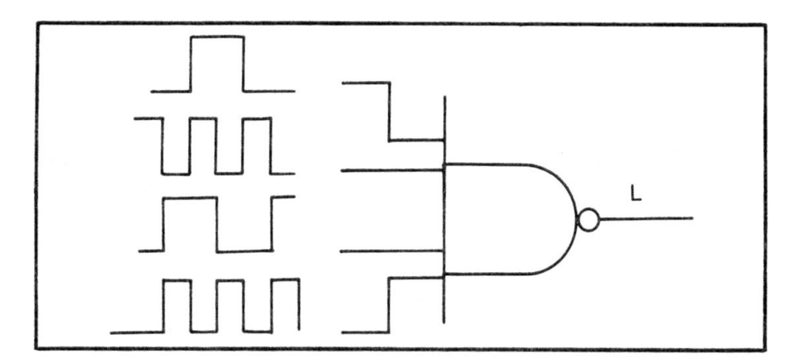

(1) a pulse. Go to Block 38.
(2) logic 0 at all times. Go to Block 21.
(3) logic 1 at all times. Go to Block 10.

BLOCK 23

Your answer to the question in Block 6 is not correct. Go back and read the question again and select another answer.

BLOCK 24

Your answer to the question in Block 9 is not correct. Go back and read the question again and select another answer.

BLOCK 25

The correct answer to the question in Block 34 is choice (1). A PAL is a programmable arrayed logic. Various logic circuits can be obtained from this integrated circuit.

Here is your next question:

Which of the following is easily toggled?

(1) R-S flip-flop. Go to Block 17.
(2) J-K flip-flop. Go to Block 36.

BLOCK 26

Your answer to the question in Block 35 is not correct. Go back and read the question again and select another answer.

BLOCK 27

Your answer to the question in Block 36 is not correct. Go back and read the question again and select another answer.

BLOCK 28

Your answer to the question in Block 1 is not correct. Go back and read the question again and select another answer.

BLOCK 29

Your answer to the question in Block 19 is not correct. Go back and read the question again and select another answer.

BLOCK 30

Your answer to the question in Block 37 is not correct. Go back and read the question again and select another answer.

BLOCK 31

Your answer to the question in Block 34 is not correct. Go back and read the question again and select another answer.

BLOCK 32

Your answer to the question in Block 36 is not correct. Go back and read the question again and select another answer.

BLOCK 33

Your answer to the question in Block 4 is not correct. Go back and read the question again and select another answer.

BLOCK 34

The correct answer to the question in Block 39 is choice (1). The question actually defines the term multiplexer.

Here is your next question:

Which of the following is used for making various logic circuits?

(1) PAL. Go to Block 25.
(2) ARRL. Go to Block 31.

BLOCK 35

The correct answer to the question in Block 10 is choice (1).

The NOR gate will have a logic 0 output at all times unless all of the inputs go to a logic 0 at the same time.

Here is your next question:

The circuit in this block can be called

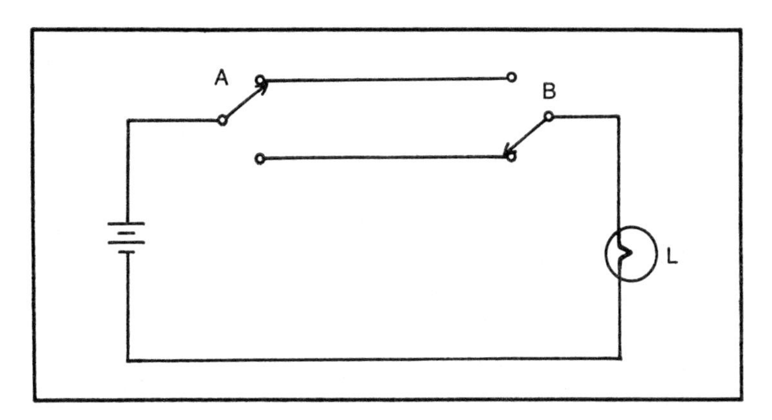

(1) INCLUSIVE OR. Go to Block 26.
(2) EXCLUSIVE OR. Go to Block 19.

BLOCK 36

The correct answer to the question in Block 25 is choice (2). You can also toggle a D-type flip-flop.

Here is your next question:

Which of the following is NOT a bus used in microprocessor systems?

(1) DATA. Go to Block 2.
(2) PROGRAM. Go to Block 40.
(3) ADDRESS. Go to Block 27.
(4) CONTROL. Go to Block 32.

BLOCK 37

The correct answer to the question in Block 19 is choice (2). This type of question has fooled many technicians. The output from the NOR is.

$$\overline{\overline{A} + \overline{B}}$$

By DeMorgan's rule, this reduces to:

$$\overline{\overline{A}}\,\overline{\overline{B}} = AB$$

Here is your next question:

Which of the circuits in this block could be used as an inverter?

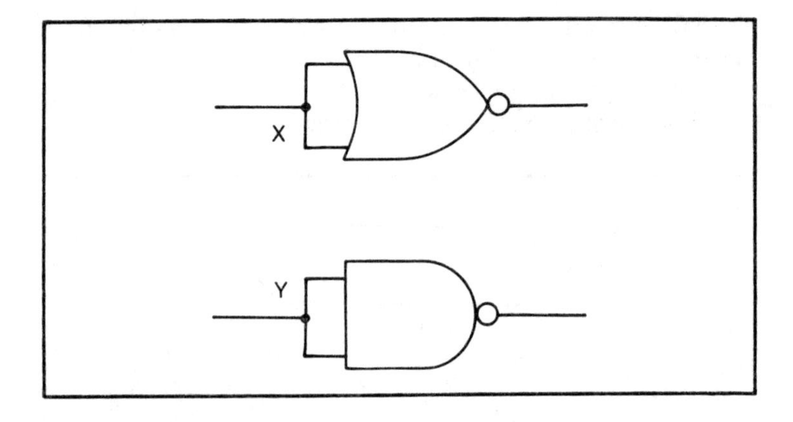

(1) Neither. Go to Block 20.
(2) Only the one marked Y. Go to Block 12.
(3) Only the one marked X. Go to Block 30.
(4) Both. Go to Block 4.

BLOCK 38

Your answer to the question in Block 22 is not correct. Go back and read the question again and select another answer.

BLOCK 39

The correct answer to the question in Block 3 is choice (1). There are two logic 0 inputs to NOR gate #1; and, there are two logic 1 inputs to NOR gate #2.

Here is your next question:

Eight different inputs can be delivered along a single data line by using a

(1) multiplexer. Go to Block 34.
(2) demultiplexer. Go to Block 16.

BLOCK 40

The correct answer to the question in Block 36 is choice (2). All microprocessors have the following three buses:

- CONTROL
- DATA
- ADDRESS

Here is your next question:

What does NMI mean on a microprocessor pin?

Go to Block 41.

BLOCK 41

The correct answer to the question in Block 40 is *non-maskable interrupt.*

You have now completed the programmed section.

ADDITIONAL PRACTICE CET TEST QUESTIONS

01. When the input to the gate of Fig. 5-10 is 010 as shown, the output at L will be.

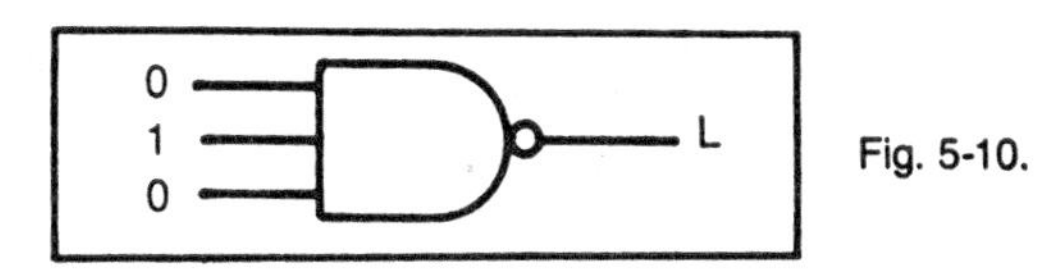

Fig. 5-10.

(1) logic 1.
(2) logic 0.

02. To change the BCD output of a counter to a signal for operating a 7-segment display, you need a

(1) ALU.
(2) 4-bit shift register.
(3) tri-state buffer.
(4) 7-segment encoder.
(5) decoder.

03. The circuit of Fig.5-11 is equivalent to

(1) a NAND.
(2) a NOR.
(3) an inverter.
(4) an OR.
(5) an AND.

04. The truth table for the circuit in Fig. 5-11 has rows marked x and y. For row x,

(1) L = 0.
(2) L = 1.
(3) cannot be determined.

05. Referring to Fig. 5-11, for the truth table of Figure 1,

(1) L = 0 for row y.
(2) L = 1 for row y.

06. Can a shift register be made with data (D) flip-flops?

(1) Yes
(2) No

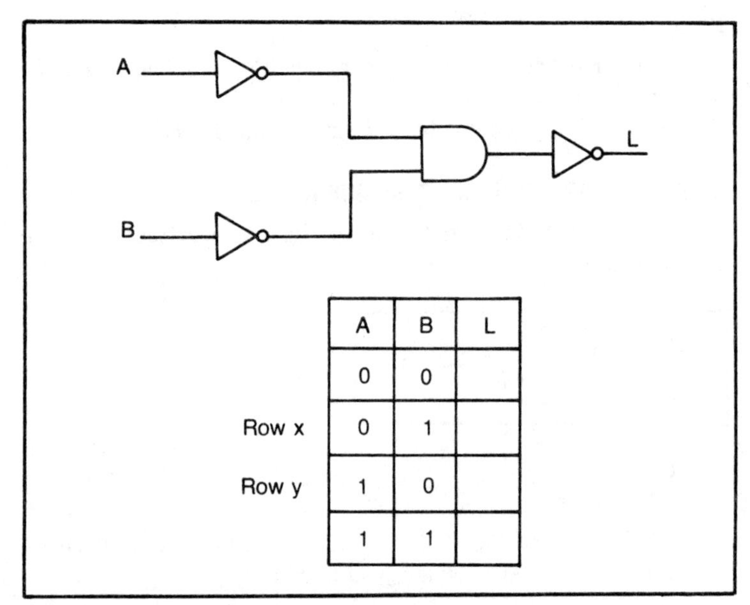

	A	B	L
	0	0	
Row x	0	1	
Row y	1	0	
	1	1	

Fig. 5-11.

07. Another name for R-S flip-flop is

(1) data latch.
(2) master-slave flip-flop.

08. The greater demand on a power supply occurs with a

(1) ripple counter.
(2) synchronous counter.

09. In order to count from decimal 0 to decimal 60, you could use

(1) a decoder.
(2) a read only memory.
(3) a random access memory.
(4) toggled flip-flops.

10. Which of the following is correct?

(1) $A + 0 = 0$
(2) $A \times 0 = 0$

11. Which of the following is an advantage of LCD over LED displays?

(1) The LCD display is much smaller.
(2) The LCD display drains the battery less.
(3) The LCD display is more rugged.

(4) The LCD display is more visible at a distance.
(5) The LCD display is more stable.

12. When there is a logic 1 delivered to input of the logic probe of Fig. 5-12, the LED marked L2 is

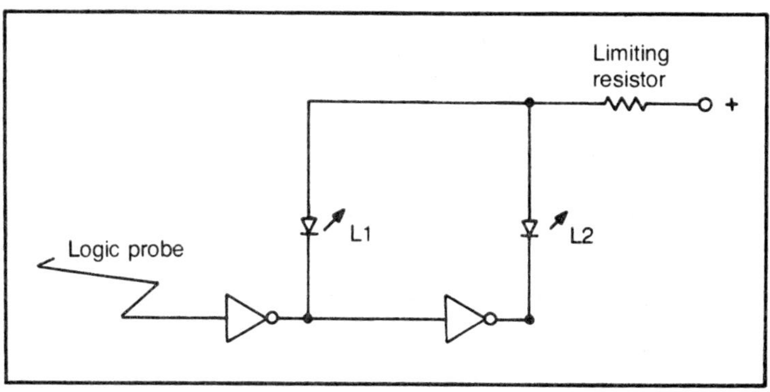

Fig. 5-12.

(1) glowing.
(2) not glowing.

13. Never connect a NAND gate output to

(1) a resistive load.
(2) logic 0.

14. A bucket brigade is used for

(1) cooling a transistor.
(2) memory.
(3) eliminating glitches.
(4) signal delay.
(5) frequency conversion.

15. After 3 input pulses at the S terminal, the output of the flip-flop in Fig. 5-13 is

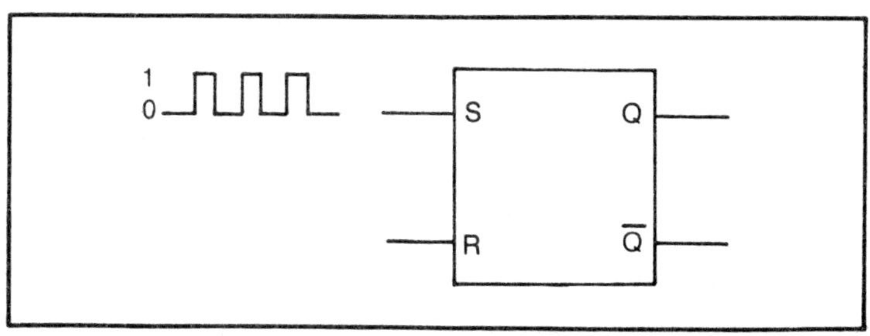

Fig. 5-13.

(1) HIGH.
(2) LOW.

16. $\overline{A} + \overline{B} = \overline{AB}$

(1) True
(2) False

17. Which of the following is the numbering base for TTL logic?

(1) 800-900
(2) 3000
(3) 7400
(4) 5000
(5) 10,000

18. Which of the following can be used for a CMOS supply?

(1) +5 V regulated
(2) −5 V regulated
(3) +15 V regulated
(4) +5 V to +15 V
(5) −5 V to −15 V

19. Compared to CMOS logic, TTL gates are

(1) faster.
(2) slower.
(3) about the same.

20. An undesired short-term voltage pulse may be caused by problems of propagation delay in combining a number of gates. The name of the undesired pulse is

(1) triffel.
(2) hitch.
(3) retch.
(4) latch.
(5) glitch.

21. The output of a tri-state buffer depends upon the

(1) signal on 1 lead.
(2) signal on 2 leads.

22. Compared to other digital families, Schottky TTL digital ICs are

(1) low speed.
(2) high speed.

23. The truth table of a simple gate describes the

(1) relation of input to output.
(2) output condition only.
(3) required for gate operation (power, ground, interconnections, etc.)
(4) input condition only.
(5) timing relationships.

24. Two types of integrated circuits are

(1) analog and amplifier.
(2) digital and logic.
(3) primary and secondary.
(4) analog and digital.
(5) (none of these choices is correct.)

25. An AND gate with two HIGH inputs will produce an output that is

(1) unstable.
(2) HIGH and then LOW.
(3) HIGH.
(4) LOW.
(5) dependent upon the previous input.

26. A NOT gate is sometimes called

(1) a switcher.
(2) a digital amplifier.
(3) a buffer amplifier.
(4) an inverter.
(5) a flip-flop.

27. Flip-flop circuits are usually toggled by an input to the

(1) Q.
(2) S.
(3) R.
(4) CLR.
(5) CLK.

28. The HIGH output of a standard flip-flop is

(1) not Q.
(2) Q.
(3) K.
(4) S.
(5) R.

29. A J-K master/slave flip-flop overcomes the problems of

(1) long-term power drain.

(2) clock input variations.
(3) unpredictable triggering.
(4) bad data storage.
(5) RESET-SET requirements.

30. Data can be placed into a bank of flip-flop registers

(1) only through the clock lead.
(2) in series or parallel.
(3) in series only.
(4) in parallel only.
(5) (none of these choices is correct.)

31. Which of the following memories is classified as volatile?

(1) PROM
(2) EPROM
(3) RAM
(4) ROM
(5) EEROM

32. A demultiplexer

(1) converts data from binary to BCD.
(2) channels several data lines to a lesser number of outputs.
(3) takes data from only one source.
(4) reduces the input voltage level.
(5) does the same thing as a flip-flop.

33. An ALU will

(1) distribute data to many inputs.
(2) distribute data to many outputs.
(3) do math, logic and data manipulation on binary numbers.
(4) do math, logic and data manipulation on decimal numbers.
(5) convert a decimal value to an ASCII code.

34. ALUs store their data in

(1) semiconductor chips.
(2) ROM memories.
(3) accumulators.
(4) shift registers.
(5) holders.

35. To translate a high-level language into machine language, you would use.

(1) a compiler.
(2) a CPU.
(3) a decoder.

(4) an encoder.
(5) an ALU.

36. A method of displaying letters, numbers, symbols, and data with binary numbers is called

(1) depletion.
(2) Boolean algebra.
(3) a program.
(4) an alpha-numeric readout.
(5) a code.

37. An internal short-term memory in a microprocessor might be called

(1) a CPU.
(2) an address.
(3) a latch.
(4) a register.
(5) a decoder.

38. When a memory is made so that data is stored in a charged or uncharged capacitor, it is called a

(1) static memory.
(2) dynamic memory.

39. Which of the following counters is faster?

(1) Synchronous.
(2) Asynchronous.

40. A certain counter is made with five flip-flops. What is the highest decimal number the counter could display?

(1) 23
(2) 31
(3) 15
(4) 464
(5) (None of these answers is correct.)

41. Which of the following is not one of the three standard buses of all microprocessors?

(1) Address bus
(2) Data bus
(3) Control bus
(4) Register bus

42. Which of the following is a method of detecting errors?

(1) Matrix code
(2) Bit slice

(3) Parity
(4) Decoding
(5) Erdet

43. The truth table shown here is for a 4-input

(1) NAND gate
(2) NOR gate.

A	B	C	D	L
0	0	0	0	1
0	0	0	1	0
0	0	1	0	0
0	0	1	1	0
0	1	0	0	0
0	1	0	1	0
0	1	1	0	0
0	1	1	1	0
1	0	0	0	0
1	0	0	1	0
1	0	1	0	0
1	0	1	1	0
1	1	0	0	0
1	1	0	1	0
1	1	1	0	0
1	1	1	1	0

44. Which of the following is mathematics used in digital studies?

(1) Boolean
(2) Pythagorean
(3) Multiposition
(4) Modern
(5) Sets

45. An advantage of parallel data handling is

(1) speed.
(2) less circuitry required.

ANSWERS TO PRACTICE TEST

Question Number	Answer Number
01	(1)
02	(5)
03	(4)
04	(2)
05	(2)
06	(1)
07	(1)
08	(2)
09	(4)
10	(2)
11	(2)
12	(2)
13	(2)
14	(4)
15	(1)
16	(2)
17	(3)
18	(4)
19	(1)
20	(5)
21	(2)
22	(2)
23	(1)
24	(4)
25	(3)
26	(4)
27	(5)
28	(2)
29	(3)
30	(2)
31	(3)
32	(3)
33	(3)
34	(3)
35	(1)
36	(4)
37	(4)
38	(2)
39	(1)

Question Number	Answer Number
40	(2)
41	(4)
42	(3)
43	(2)
44	(1)
45	(1)

Linear Circuits in Consumer Products

Linear circuits are also referred to as *analog* circuits. A simple definition of a linear circuit is that it *has an output at all times during which there is an input.* Another way of saying that is there will always be an output signal when there is an input signal. Furthermore, the output signal amplitude is continually variable from its minimum to its maximum value as the input signal amplitude varies from minimum to maximum.

It is very difficult to make a sharp distinction between linear and digital circuits. Consider, for example, the class-C amplifier. With a sine-wave input, the output of a class-C amplifier is a series of pulses. (Technically, they are partial sinusoids.) With this type of amplifier you cannot say there is always an output signal when there is an input. Nevertheless, class-C amplifiers are considered to be linear. A real distinction between linear and digital is that a digital circuit normally operates between two fixed values. It has no stable condition in between those two values.

Before you take the CET Test, you should review the characteristics of class-A, B and C amplifiers. Also, review AB1 and AB2 amplifiers. Important things to remember are that the linear class-A amplifier has the least amount of output distortion, whereas the class-C amplifier has maximum distortion. However, the class-C amplifier has the maximum efficiency. Some of the characteristics of class-A, class-B, and class-C amplifiers were discussed earlier in this book.

There is a considerable emphasis in the newer CET Tests on integrated circuit linear devices. Of course, one of the most popular of these is the *Operational Amplifier.* That will be the first subject reviewed in this chapter.

INTEGRATED CIRCUIT OPERATIONAL AMPLIFIERS

An *operator* in mathematics is a mathematical symbol (such as $+$, $-$, $\div$, and $\times$) that tells the person working the problem what he is supposed to do. For example, the plus sign tells the person to ***add*** the second number to the first number.

When you studied basic electricity and electronics, you probably studied about *j* operators. If you learned that subject correctly, you understood that *j* was a mathematical operator just like the plus and minus signs in arithmetic A $+j$ tells you to make a ***left turn*** when moving down the X axis; and, A $-j$ simply tells you to make a ***right turn*** when you are coming down the X axis.

In the late 1930s and during World War II, *analog computers* were the most important type of computer available. With an analog computer, the input data is continuous, and it is operated upon by very specialized amplifiers to obtain the desired result.

Suppose, for example, you needed to add two numbers in an analog computer. The two are delivered to the computer as reference voltages. So, in a simple example, if you wanted to add the number 3 to the number 2, you could represent the 3 with a 3-volt input, and represent the number 2 with a 2-volt input. These two inputs were applied to an amplifier where they were combined to produce an output voltage of 5 volts. The result means that $3 + 2 = 5$.

It didn't take the designers very long to realize that any time they wanted to perform some kind of analog mathematical operation, the starting point was the amplifier. Why go to the trouble of designing a new amplifier every time you wanted to solve a new problem? It would be much better (and this was the way it was accomplished) to design one amplifier that could be modified for a wide variety of mathematical operations. That became the operational amplifier.

It can be proved with a mathematical analysis that any time a high-gain amplifier utilizes a negative feedback, *the output of that amplifier can be made dependent only upon the feedback network*! In order to accomplish this, it is necessary that the amplifier have one important characteristic—that is, it must have a *very high open-loop gain*. (The open-loop gain of the amplifier is the gain of the amplifier without the negative feedback network.)

I have advertisements in my files from the middle 1940s that brag about vacuum-tube operational amplifiers having gains as high as twenty and thirty thousand. Simply speaking, the output would be 20,000 or 30,000 times greater than the input when you use these amplifiers in an open-loop configuration. The integrated circuit operational amplifiers used today have gains in the range of 500,000 to well over 3 million!

The second characteristic of an operational amplifier is that is must have a *linear rolloff*. This is necessary so that the bandwidth and the gain tradeoff is predictable. Figure 6-1 shows a typical *bode plot* of an operational amplifier. This curve is for an integrated circuit type that is identified by the number 741. The 741 op amp has become a standard by which other operational amplifiers are compared.

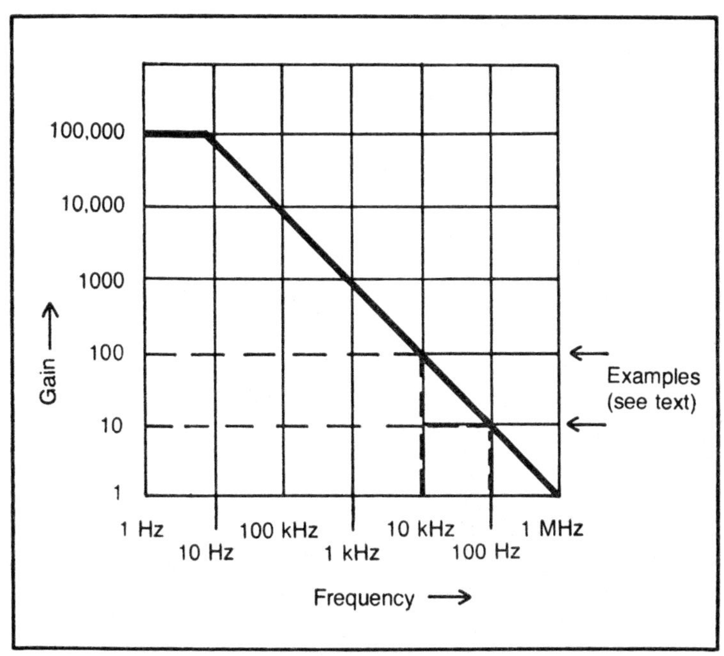

Fig. 6-1. This Bode plot shows the linear rolloff for a 741 operational amplifier. The tradeoff between gain and bandwidth is shown by the examples.

Historically, the 741 was not the first integrated circuit operational amplifier to be introduced to the market. It was preceded by the number 709. That op amp had most of the same characteristics as the 741 (although the 741 was an improvement). However, the

709 had a very undesirable characteristic. If the input signal exceeded a certain predetermined amplitude, it would *latch up*. That simply means that the output went to its maximum value and stayed there. The only way you could get it out of a latch up condition was to turn the system off.

It is a relatively simple matter to add external circuitry to the 709 to prevent it from latching up, but that means an additional cost and an additional space requirement. Those are disadvantages that the design engineers were not willing to put up with.

The bode plot for the op amp is an excellent example of the tradeoff between gain and bandwidth. Two examples are marked in dotted lines on the bode plot. Notice that when the gain is 10 the bandwidth is 100 kHz. However, when the gain is increased to 100, the bandwidth decreases to 10 kHz. Therefore, increasing the gain in the op amp reduces the bandwidth. This is the tradeoff of all amplifiers that were discussed earlier in the book.

Typical CET Test Question

An operational amplifier is being used in a system that requires a bandwidth of 1000 Hertz. The Bode plot of the op amp is shown in Fig. 6-1. What is the maximum gain of this operational amplifier?

(1) $A_v = 100$
(2) $A_v = 1000$
(3) $A_v = 10{,}000$

Answer: (2) When the operational amplifier has a gain of 1000 (A_v = 1000) the bandwidth is 1 kilohertz as shown on the Bode plot.

Although all amplifiers have a gain-bandwidth tradeoff, there is no direct tradeoff between those parameters unless there is a *linear rolloff* as shown in this bode plot.

A third characteristic of the operational amplifier is that it must have a *differential input*. This simply means that a differential amplifier is used at the input stage. It is assumed when you take the CET Test that you understand how a differential amplifier works. It was discussed briefly in a previous chapter.

There is an important thing about the differential input that you must understand. *A common mode operation exists when you connect the two inputs together*. This is known as a common mode connection for the input terminals. Theoretically, if the differential amplifier is perfectly balanced and a signal is applied to the two inputs simultaneously the output should be zero. That will be true only if it is a perfect differential amplifier.

If any unbalance exists in the two differential amplifier sections, then a common-mode input will produce a slight output signal. Remember, in reality there is no such thing as a perfect differential amplifier.

Many circuits are available for operational amplifiers to set the balance between the two input sections of the differential amplifier. To understand why that is necessary, suppose that one of the amplifiers in Fig. 6-2 is conducting harder than the other. This is in the absence of any signal. The unbalance is highly undesirable because there would always be an output voltage even though there is no input voltage.

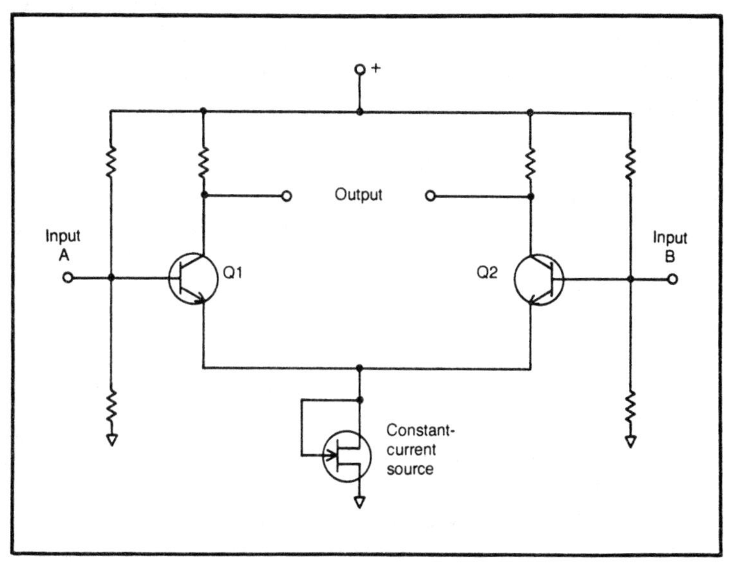

Fig. 6-2. A differential amplifier circuit.

To eliminate that possibility, a dc bias is applied to the two amplifiers. The bias voltage is adjusted so that the output of the two amplifiers is identical. With this *offset* bias adjustment being made, the output of the operational amplifier can be made very nearly equal to zero in the absence of an input signal—especially in the common mode connection.

Typical CET Test Question

To make a common-mode measurement

(1) the input terminals are connected together.

(2) the output terminals are connected together.

Answer: (1) When the input terminals are connected together, and an a-c signal applied to the common connection, the output should be zero volts. This is a common-mode rejection test.

What you would like to have in an ideal operational amplifier is an enormously high open-loop gain and a very, very small common-mode output. So, when you go to buy an operational amplifier, you can look in the specifications to find one specification called the CMRR, or *common mode rejection ratio*. That's simply a specification that tells how closely you approach the ideal condition just mentioned.

A fourth characteristic of the operational amplifiers, is that they must have a *high input impedance* and a *very low output impedance*. Why is that important? Well, in the first place, the high input impedance is needed so that the operational amplifier does not load the signal source. In many applications, the signal source is not able to deliver any appreciable amount of load current.

The output impedance should be nearly zero ohms so it can be connected to a low-impedance device without having an undesirable mismatch. In the 741 operational amplifier just mentioned, the output impedance is so low that when it is operating, you can short-circuit the output terminal to ground without having a disastrous overload current.

In order to achieve an output impedance that's very close to zero ohms, the output circuit of most operational amplifiers is in a totem pole form. With long-tail bias, the totem pole operates with a positive and negative supply. If you go halfway between the positive and negative voltages, you're at zero volts and also nearly zero impedance. It should also be pointed out here that long-tail bias is used for the differential amplifiers at the input of most operational amplifiers. In review, long-tail bias means that the amplifier is connected between a positive and negative voltage source.

One other characteristic of the operational amplifier should be understood. It is called the *slewing rate*. This is simply a method of evaluating an operational amplifier as to how fast it can change its output voltage with an input step function. In other words, if the input voltage goes from zero to some value instantly, the output should (theoretically at least) also go to its maximum value instantly. No operational amplifier can actually achieve this.

The time that it takes for the output voltage to go from zero to its maximum value with a step input is the slewing rate. It is usually measured in volts per microsecond, or some similar rating. If you are going to use an operational amplifier in a high-frequency situation, you will certainly want a high slewing rate.

It is not uncommon to use operational amplifiers in digital circuits. For example, you can wire an operational amplifier to produce a pulse or square-wave output. In order for the pulse to be applicable to any digital system, it must have a very fast rise time and a very fast decay time. Remember this important thing about digital circuits: The counters and many of the other circuits in a digital system *operate only during the rise time or the decay time of the input pulse*. If you have a long rise time or a long decay time, you will not get a fast switching response, and that is undesirable in most digital systems. To produce such a pulse with an operational amplifier, it must have a high-valued slewing rate.

Having looked at the operational amplifier and some of its specifications, it is now time to look at a few basic circuits. The following section is, in no way, intended to represent the total range of applications of operational amplifiers. Even listing those without accompanying circuits would take more room than this complete chapter is devoted to. The examples given in this chapter are typical, and ones that you are very likely to see with related questions in a CET Test.

The Inverting Amplifier. The term *inverting amplifier* simply means that the input signal is shifted 180° in phase by the operational amplifier. The circuit for the operational amplifier connected as an inverting amplifier is shown in Fig. 6-3. The gain of this simple circuit can be calculated very simply by the following equation:

$$A_v = -\frac{R_f}{R_i}$$

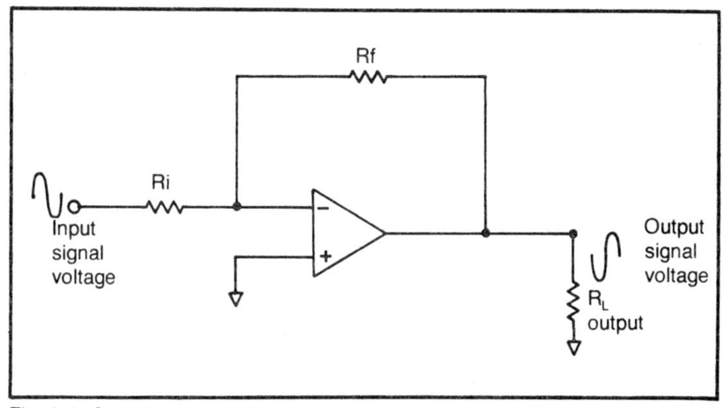

Fig. 6-3. Operational amplifier in the inverting configuration.

Typical CET Test Question

The negative sign in the voltage gain equation for the circuit of Fig. 6-3 means

(1) the output signal is negative at all times with respect to common.

(2) the output signal is 180° out of phase with the input signal.

Answer: (2)

Note that the gain in the above equation is dependent only upon the feedback and input resistance circuitry, and it has nothing to do with the characteristics of the amplifier itself. As mentioned before, if the amplifier has a sufficiently high gain, then this type of gain equation will be accurate.

Technicians have an interesting way of explaining why the op amp output is dependent upon this feedback ratio. To begin with, the input resistance must be relatively high (usually never less than 1000 ohms) in order to prevent loading of the input circuit. The junction between the input resistance and feedback resistance is at the inverting terminal. This point is technically referred to as the *summing point.*

If the operational amplifier is operating properly, the voltage at this point should be zero volts. In the simplified explanation, the operational amplifier performs its job by maintaining the summing point at zero volts. Another way of saying this is that *the operational amplifiers always performs in such a way as to make this voltage zero volts.*

If the feedback resistor (R_f) is a high value, then the gain of the operational amplifier is also high. This is necessary so that *when the output signal is dropped across the feedback resistor, the voltage will be zero volts at the summing point.* Conversely, if the feedback resistor is very small, then the gain of the operational amplifier does not have to be very high in order to produce enough signal through the feedback resistor to maintain the summing point terminal at zero volts.

The Noninverting Amplifier. The noninverting amplifier connection is shown in Fig. 6-4. Observe in this case that the feedback network is still in place, and the gain of the amplifier will still be dependent upon that feedback network. This is necessary in order to achieve zero volts at the summing point. The gain of the circuit in Fig. 6-4 is given as follows:

$$A_v = 1 + \frac{R_f}{R_i}$$

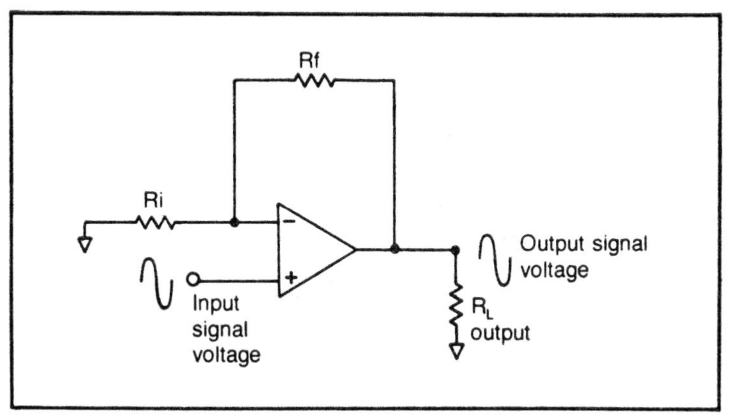

Fig. 6-4. This noninverting configuration has a slightly higher gain than the inverting amplifier of Fig. 6-3.

It is obvious that the gain of the noninverting amplifier is very slightly higher than for the inverting amplifier. However, if the ratio of the feedback resistance to the input resistance is high (meaning that the gain of the amplifier in a closed-loop configuration is high), then the gain of the inverting and the non-inverting amplifier are virtually the same values. Note that the output signal is in phase with the input signal in this noninverting configuration.

The Operational Amplifier As A Follower. If you think back to the discussion in an earlier chapter about follower circuits, you will remember that one important characteristic they have is a high input impedance and a low output impedance. The second characteristic is that the follower does not invert the phase of the input signal. These two characteristics make it very simple to connect an operational amplifier as a follower. This configuration is often referred to as a *buffer*, which is simply an amplifier that has a gain of 1 and is connected between two circuits to match impedance or to isolate one circuit from another.

Figure 6-5 shows the circuitry for the op-amp buffer. Notice that there is no feedback resistance and no input resistance in this particular application. However, this same gain of 1 could also be accomplished by using a feedback resistance of 2500 ohms and an input resistance of 2500 ohms. The configuration shown in Fig. 6-6 is *not* preferred. (You will remember that the gain of a noninverting amplifier is very slightly higher than the ratio of the feedback resistances, so the feedback circuit would have to be trimmed in order to get a gain of 1. This is an unnecessary bother and that explains one reason why the circuit of Fig. 6-5 is preferred.)

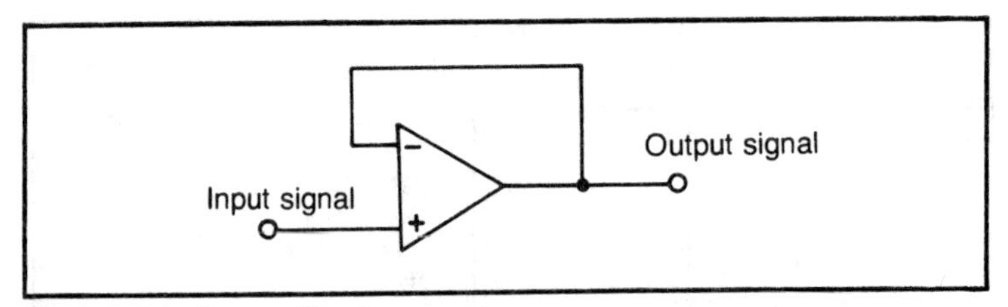

Fig. 6-5. A buffer amplifier has no gain.

Additional Op-Amp Applications. There are some additional applications of operational amplifiers that you should be familiar with. They will be described here, but the individual circuitry will not be discussed. As an experienced technician, you should make yourself familiar with this type of technology because you are likely to see applications in consumer products.

Gyrator—A gyrator is an operational amplifer circuit that is connected in such a way that it behaves like an inductor.

Comparator—Because of the differential amplifier input, an operational amplifier is ideally suited for making a comparison between two signals. Specifically, it can compare two signals to see which of the two is more positive.

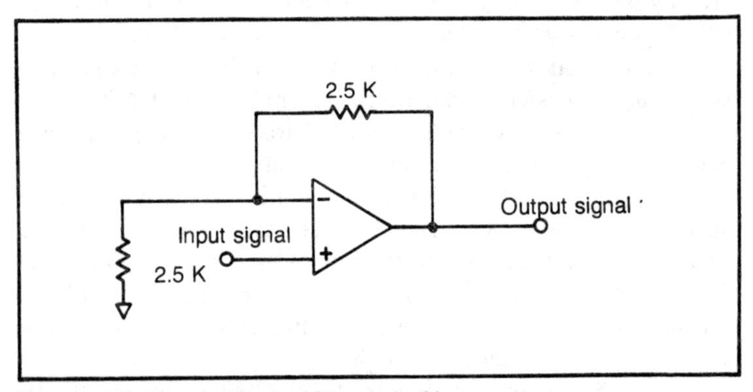

Fig. 6-6. A buffer amplifier *could* be made this way.

Timers—Timers are linear integrated circuits that can be used to delay the start of a signal or system. These very versatile timers are also utilized as oscillators, sensors and other circuits. The 555 timer is one of the most popular of these. It employs two operational amplifiers in *voltage comparator* and *shut down* circuitry.

Compander and expander—The dynamic range of any signal is the maximum range of values between its minimum and maximum amplitude. If the dynamic range of a particular signal is greater than the dynamic range of the circuit to which it is to be delivered, it can be passed through a *compander*, which preserves the relative voltages between the different parts of the signal, but at the same time, decreases the total dynamic range. By contrast, an *expander* takes a relatively narrow range of signal voltages and increases it to a large dynamic range.

Sample and hold—Sample and hold circuits are usually constructed with two operational amplifiers. They are especially useful in analog-to-digital (A/D) converters, and in measurement circuits. Specifically, at the request of an input signal, the sample and hold circuit maintains an output that is proportional for the amplitude of the input signal.

Active Filter—A passive filter uses only resistors, capacitors, and inductors. Active filters use amplifiers to obtain a desired input and output signal characteristic. Operational amplifiers are used extensively in active filter circuits.

There are two other integrated circuits which will be discussed briefly in this chapter. They are also examples of the linear type. One is the phase-locked loop (to be discussed next), and the second is the analog power supply regulator.

PHASE-LOCKED LOOP

Some early television receivers used phase-locked loops to maintain the horizontal oscillator on frequency. One example was the synchrolock type of automatic frequency control. This vacuum tube circuit, and others like it, employed a closed-loop system that set the oscillator frequency equal to the horizontal synchronizing pulses coming from the transmitter.

Figure 6-7 shows a block diagram of a modern system that is produced in integrated circuit form. An input signal is compared with the frequency of an oscillator. If those frequencies are equal in frequency and phase, then there is no output correction voltage.

Any difference in frequency or phase results in an output voltage. That voltage passes through a low-pass filter. Its purpose is to assure that any oscillator or incoming signal does not pass. The correction voltage must be a pure dc.

A dc amplifier follows the filter. It converts small changes in the dc correction voltage to larger voltage changes. The oscillator frequency is controlled by a varactor diode. If its frequency does not exactly equal the incoming frequency, then the correction voltage from the dc amplifier adjusts the varactor capacity until there is a frequency and phase match.

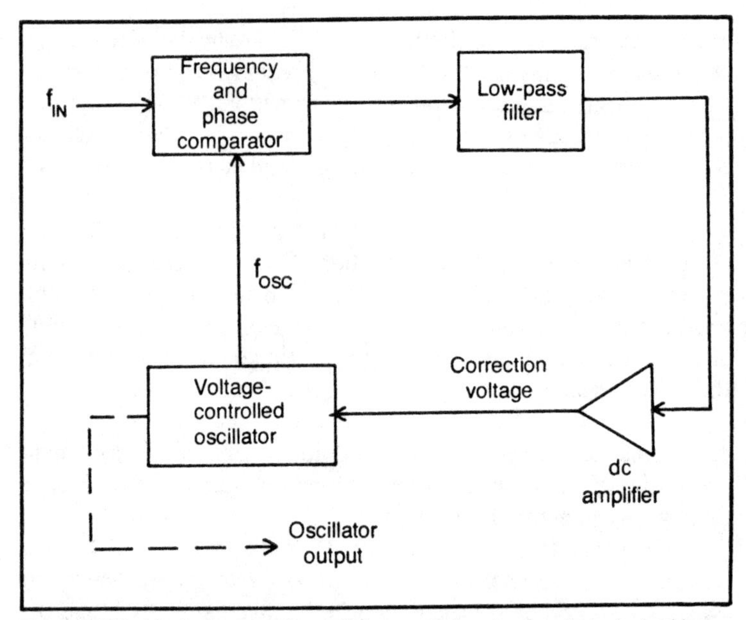

Fig. 6-7. This block diagram shows the essential parts of a phase-locked loop.

One application of the circuit is to obtain an oscillator output frequency that is locked to the incoming frequency. The required output signal is delivered from the VCO as indicated by the broken arrow.

If the incoming signal is frequency modulated, then the correction voltage will continuously try to correct the oscillator. The result is that the correction voltage will follow the modulating signal for the FM input signal. This means that the correction voltage is actually the same signal that modulated the FM waveform. The overall result is that the circuit will behave like an FM demodulator, and the audio output signal is taken from the output of the dc amplifier.

The phase-locked loop can also serve as a frequency synthesizer. One application is to obtain a local oscillator frequency that is always an exact multiple of the incoming frequency. Figure 6-8 shows the basic circuit. A programmable counter has been added to the basic circuit. It divides the VCO frequency by an amount equal to N. This value is set by the microprocessor.

Keep in mind the fact that the input to the comparator must be equal in phase and frequency to the incoming signal (f). Therefore, the oscillator frequency must be N × f so that when it is divided by N the signal to the comparator will equal f:

$$\frac{f \times N}{N} = f$$

You will find the system of Fig. 6-8 in tuners of TV receivers, and in the rf section of communications receivers.

Typical CET Test Question

For the system shown in Fig. 6-8, the input frequency (f_{IN}) is 200 kilohertz. To get an oscillator frequency of 400 kilohertz the ÷ N counter must be set to

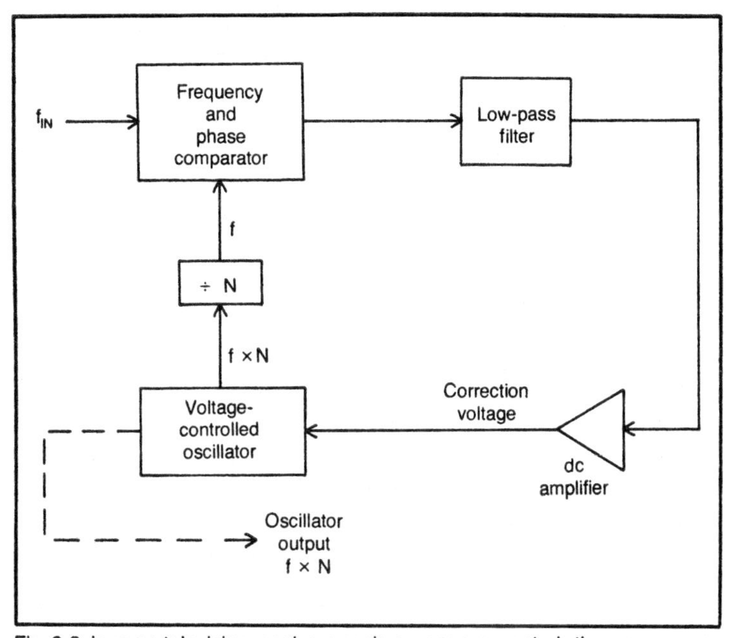

Fig. 6-8. In some television receivers a microprocessor controls the programmable counter (÷ N).

(1) ½
(2) 2

Answer: (2) If the oscillator operates at 400 kilohertz, and its frequency is divided by 2, the output of the ÷ N programmable counter will be 200 kilohertz. A phase lock occurs because that frequency equals f_{IN}.

ADDITIONAL ANALOG CIRCUITS

Dolby Audio Systems. Dolby systems are used in audio to

reduce noise. Dolby noise reduction systems can now be purchased in integrated circuit form.

Tracking Regulators. Tracking regulators are used when there is both a positive and a negative dc voltage out from a power supply. Using the tracking regulator configuration, any change in the positive voltage setting will be automatically accompanied by an identical change in the negative output voltage. As an example, if the output voltage is increased from 10 to 12 volts (by adjustment from an operator) then the negative output voltage will go from −10 to −12 volts automatically.

Oscillator. Oscillators are classified in two groups: *sinusoidal* and *nonsinusoidal* (or relaxation). Both kinds are readily available in integrated circuit form. As a general rule, the timing capacitors or inductors for IC oscillators are located externally to the integrated circuit package.

Remember this important definition of an *oscillator*: it is a circuit that converts dc to ac. Circuits which convert ac to dc are called *rectifiers*. *Converters* change dc from one value to another, while *inverters* change dc to ac. Since inverters and oscillators both change dc to ac, you can presume that most inverters are forms of oscillator circuits.

AMPLIFIER COUPLING CIRCUITS

As a technician you should be thoroughly familiar with the methods of coupling a signal from one amplifier to another. When the output signal of one amplifier feeds into the input of another amplifier, the combination is said to be *cascaded*. Do not confuse this term with *cascoded* which is a special type of rf amplifier.

When the output of one amplifier is connected to the input of the next amplifier with a piece of wire, the two amplifiers are said to be *direct coupled*. In the olden days, when they were made with vacuum tubes, the term Loftin-White amplifier was used to refer to these direct-coupled circuits.

The obvious advantage of direct coupling is that it has a very high frequency response. This is achieved by the fact that there is no reactive component in the coupling circuit. There is, however, reactance at the input of the one amplifier and from the output of the other amplifier. This is usually capacitive reactance and it has the effect of limiting the high-frequency response of direct coupling.

Resistor-capacitor coupling (which is often called RC coupling) is the least expensive method of coupling two amplifiers. You might think that direct coupling would be simpler and cheaper, but there is a problem of *level shifting* in direct coupling. Usually, the input of one amplifier is at a different dc level than the output of the other, and so, the second amplifier must often be operated at a higher voltage. Unless an interfacing circuit is used, *the power supply must be more elaborately designed*. This is not only true because of the different dc levels, but remember if there is any ripple on the power supply, that ripple will be delivered to the second amplifier and will be treated as an ordinary signal. In other words, *the ripple will be amplified* considerably.

RC coupling has a limited low frequency response because the coupling capacitor has a high reactance to low frequencies. To circumvent the problem of low-frequency response, some amplifiers will use a low-frequency compensating network. This is illustrated in Fig. 6-9. The principle is simple to understand.

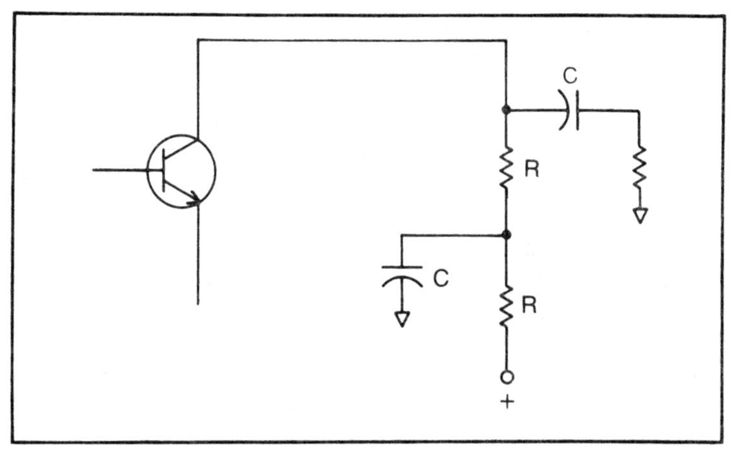

Fig. 6-9. This low-frequncy compensating network has the same layout as a decoupling filter.

At low frequencies the capacitance at the junction of the two resistors in the collector circuit is virtually an open circuit. In other words, it has a high reactance. Therefore, at low frequencies the gain of the amplifier is set by a high value of collector load resistance consisting of the two resistors in series.

Even though the low-frequency signal is amplified with more gain, the actual signal that arrives at the next amplifier is still not greatly increased in amplitude because of the increased reactance of the coupling capacitor. So, the gain is increased to overcome the high loss in the coupling capacitor at *low frequencies*.

At high frequencies the coupling capacitor becomes a short

circuit (virtually) so the signal is coupled from one amplifier to the next with very little loss. At the same time, the capacitor at the junction of the collector resistors also becomes a short circuit. For all practical purposes, the lower resistor is shorted out, and the collector load resistance is reduced to one resistor. The reduced collector resistance reduces the gain of the stage, but less gain is needed because the loss due to the coupling capacitor is eliminated.

This low-frequency compensating network looks identical to a decoupling filter for a power supply line. You can tell which is which by the value of the resistance in the circuit, and also by the value of the capacitor. Specifically, the resistor must be relatively large if it's a low-frequency compensating network. Otherwise, it would not materially add to the collector load resistance when the low frequencies are being boosted. On the other hand, if it's a decoupling filter, a resistor will be very small (10 ohms is common). The reason is that it is not desired to drop any appreciable amount of dc power supply voltage across the decoupling filter.

Another kind of coupling is obtained with a transformer. *Transformer coupling* has an important advantage: *it can be tuned or used to select some particular frequency or band of frequencies*. This is especially true when the transformer between stages is an air core or ferrite type. That situation is important to a technician. By looking at the transformer coupling between two stages, he can determine whether this is an audio circuit (using an iron-core transformer) or an rf circuit (using a tuned air-core transformer).

Transformer coupling also has the advantage of being broadband, but has the disadvantage of being frequency selective in broadband circuits. Don't forget the transformer coupling is useful in phase-splitting applications. So, when a phase inverter is needed for push-pull operation, a transformer can easily supply the 180° out-of-phase signals.

Impedance coupling between amplifiers is obtained by combinations of resistance, inductance, and capacitance. Impedance coupling may, in later designs, use a tunable coil so it can pass a specific band of frequencies. You will see that type of circuit used in intermediate-frequency amplifiers in television systems. They replace the transformer coupling which has the disadvantage that it was more difficult to obtain a complete 6-megahertz bandwidth over the complete range of i-f frequencies.

POWER SUPPLY CIRCUITS

Since a great amount of effort has been made to avoid using

trade-name circuits that are not familiar to all technicians, you can expect circuits used in the CET Test to be of the type which are familiar to all types of technicians. One circuit that should be familiar to everybody that works in electronics—regardless of what system brand name they're working on—is the power supply. Battery-type power supplies are not generally considered, so what you are likely to encounter is a rectifier-type supply.

An important thing to do is to become familiar with the various rectifier configurations. You should be able to recognize *half wave*, *full wave*, *bridge rectifiers*, *half-wave doublers*, and *full-wave doublers*. Discussion on these basic rectifier circuits will not be given here. If you encounter them in a Journeyman CET Test, it will most likely be in conjunction with some other circuitry.

It is useful to understand the term regulation as it applies to an unregulated supply. Regulation is a measure of how well the supply holds its output voltage under varying load conditions. The equation for percent regulation of a power supply is given by

$$\%\text{ Regulation} = \frac{\text{No-Load Voltage} - \text{Full-Load Voltage}}{\text{Full-Load Voltage}} \times 100$$

The term *percent regulation* is almost meaningless in a regulated supply. If the unregulated circuit is properly designed, it will not be possible to get a measurable change in output voltage for changes in load current over the specified values called out by the manufacturer designer.

Figure 6-10 shows an example of a simple *closed-loop analog*, or, *continuous-voltage regulator*. This circuit can be divided into some basic functions. The sense circuit, which is comprised of three resistors in series, is used to sense the output voltage of the power supply. In order for regulation to take place, the regulator circuit must know what the output voltage is, and the divider network provides this information.

The second thing that is necessary in a regulated circuit is a reference voltage. This is usually obtained with a zener diode. In the sense amplifier, the reference (or zener) voltage is compared with the sense voltage of the power supply. The way the circuit is designed, these two voltages must have a specific ratio. If the ratio is too high, then the amplifier will deliver a signal to the series-pass transistor which lowers the power supply voltage.

Note that the resistors for the voltage sense circuit in the circuit of Fig. 6-10 is connected across the output voltage, which is

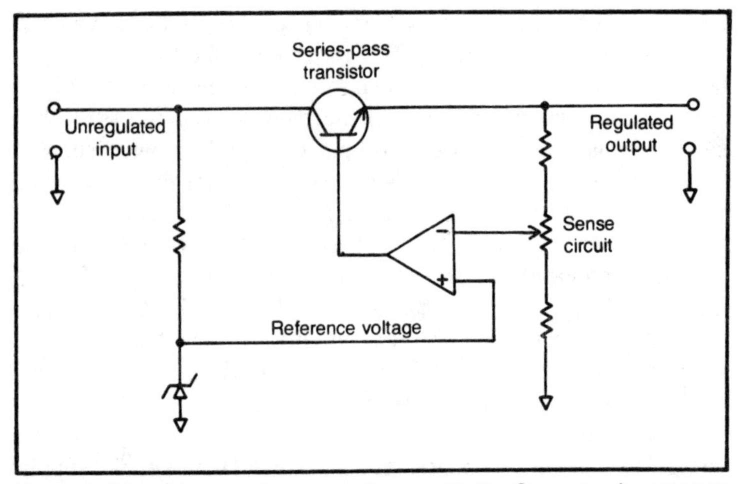

Fig. 6-10. Closed loop regulators are often used in the Consumer Journeyman CET test.

what you would expect. If a *current sense* is needed, *the sense circuit is connected in series with the power supply load.*

Figure 6-11 shows a *current limiter* with a series sense resistor. If the voltage across this resistor is above 0.7 volts, the silicon transistor conducts. Its collector current flows through R, reducing conduction through the series-pass transistor. (Remember that the term *load* in a power supply always refers to the current in the output. Sometimes technicians confuse the term *load* with *load resistance.* The word load resistance represents the opposition, or the useful output of a power supply, but the load is always the power supply current.)

Instead of using an analog feedback system like the one shown here, it is possible to have a regulated power supply which uses a

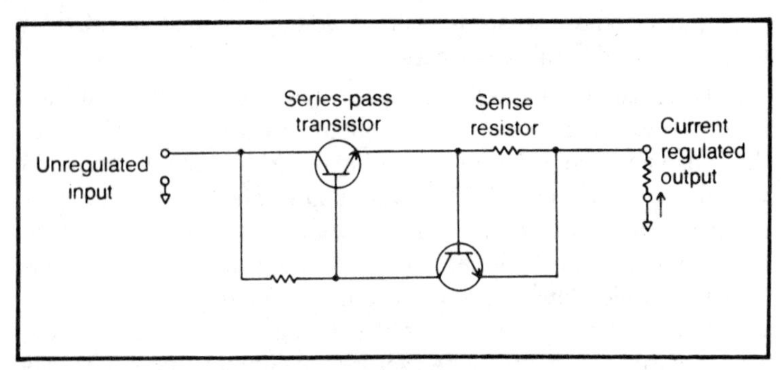

Fig. 6-11. Power supply *current* is controlled by this regulator.

digital signal. This type of supply is usually referred to as a *switching regulator.*

Switching regulators continually turn the load current on and off. The amount of ON time is determined by the closed-loop regulating circuitry. If the output voltage or power is too low, then the load current is switched on for a longer period of time. Conversely, it is switched on for a shorter period of time if the output voltage or load power is too high. By regulating the ON time, it is possible to control the output voltage and power of the supply. Switching regulators are more efficient than the analog type.

A word should be said about troubleshooting closed-loop power supply circuits. The same technique for troubleshooting these circuits is used for troubleshooting any closed-loop system. *The best way to do this is to open the loop at some point and substitute a fixed voltage.* By opening the loop, you eliminate the feedback circuitry and the erroneous measurements that you are liable to get in a closed-loop system that is not working properly.

Another type of power supply you should be familiar with is *scan-derived.* It obtains the ac voltage necessary for obtaining output dc power from the flyback transformer. Scan-derived power supplies are very popular in newer television receivers. Since the output dc voltage from this supply is dependent upon a signal from the flyback transformer, and the signal from the flyback transformer comes from the amplifiers that require a dc voltage, it is obvious that this system cannot start by itself. A special starting circuit is necessary to start the horizontal oscillator in the flyback system in order to start an ac voltage which can be rectified and utilized in the system.

A special circuit of interest is shown in Fig. 6-12. It is called a *crowbar.* It consists of an SCR connected across the load in a power supply system. It is a protective circuit and is used to prevent integrated circuits, and other voltage-sensitive circuits, from a temporary or permanent overvoltage.

If an overvoltage occurs, a positive-going gate signal is delivered to the SCR which short circuits the output terminals. The term crowbar comes from the fact that the power supply behaves as if a crowbar was connected across its output terminal when it is shut down by the SCR.

Typical CET Test Question

In a certain television receiver the ac input voltage to the rectifier comes from the flyback transformer. This is an example of

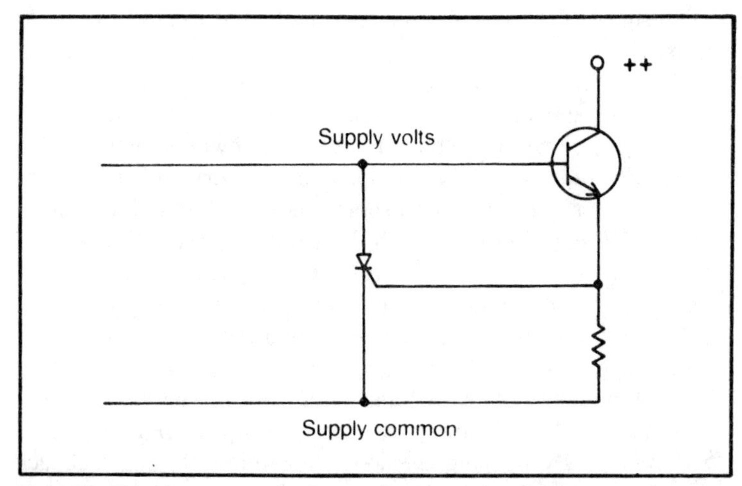

Fig. 6-12. This is one example of a crowbar circuit. It protects components that are connected to the power supply output from an overvoltage.

(1) a switching regulator.
(2) a scan-derived supply.

Answer: (2)

PROGRAMMED REVIEW

Start with block number 1. Pick the answer that you feel is correct. If you select choice number 1, go to Block 13. If you select choice number 2, go to Block 15. Proceed as directed. There is only one correct answer for each question.

BLOCK 1

The voltage gain of a common-collector circuit is

(1) less than unity. Go to Block 13.
(2) moderately high. Go to Block 15.

BLOCK 2

Your answer to the question in Block 37 is not correct. Go back and read the question again and select another answer.

BLOCK 3

The correct answer to the question in Block 19 is choice (2). The operational amplifier is connected in a noninverting configuration.

Here is your next question:

The rectifier circuit in this block is called a

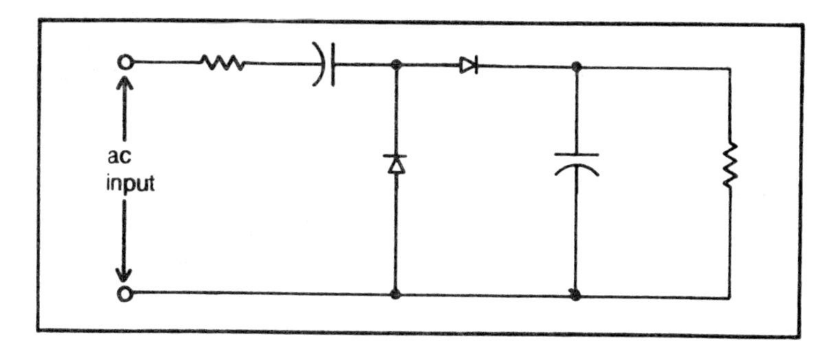

(1) half-wave doubler. Go to Block 30.
(2) full-wave doubler. Go to Block 14.

BLOCK 4

The correct answer to the question in Block 38 is choice (1). Slewing rate tells how fast the output can change with an input step function.

Here is your next question:

It would be best for an operational amplifier to have

(1) a high value for its common-mode rejection ratio. Go to Block 31.
(2) a low value for its common-mode rejection ratio. Go to Block 26.

BLOCK 5

Your answer to the question in Block 31 is not correct. Go back and read the question again and select another answer.

BLOCK 6

The correct answer to the question in Block 36 is choice (1). Remember that the output of a differentiator or integrator has a sinusoidal waveshape when the input signal is a sine wave.

Here is your next question:

Increasing the resistance of the collector load resistor in a transistor amplifier will

(1) decrease its gain. Go to Block 28.
(2) increase its gain. Go to Block 35.

BLOCK 7

Your answer to the question in Block 27 is not correct. Go back and read the question again and select another answer.

BLOCK 8

Your answer to the question in Block 35 is not correct. Go back and read the question again and select another answer.

BLOCK 9

Your answer to the question in Block 18 is not correct. Go back and read the question again and select another answer.

BLOCK 10

Your answer to the question in Block 11 is not correct. Go back and read the question again and select another answer.

BLOCK 11

The correct answer to the question in Block 24 is choice (1). Moving the arm of the variable resistor all the way to point x will short out the capacitor. Then, the high frequencies will be shunted around the speaker. Since the high-frequency response of the amplifier is poor, the low frequencies will be more prevalent. The circuit is called false bass because the bass response of the amplifier is not really improved.

Here is your next question:

In the parallel-resonant circuit shown in this block, moving the plates of the capacitor closer together will

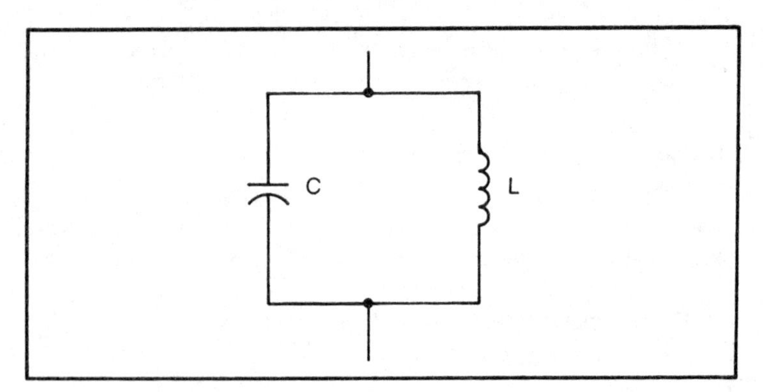

(1) raise the resonant frequency. Go to Block 10.
(2) lower the resonant frequency. Go to Block 36.

BLOCK 12

The correct answer to the question in Block 30 is choice (2).

The operational amplifier is connected in an inverting configuration, so the voltage at point y is 180° out of phase with the voltage at point x. Also, you want to *decrease* the conduction in the series-pass transistor when the output voltage is increasing.

Here is your next question:

In the amplifier circuit shown in this block, the voltage at point x should be

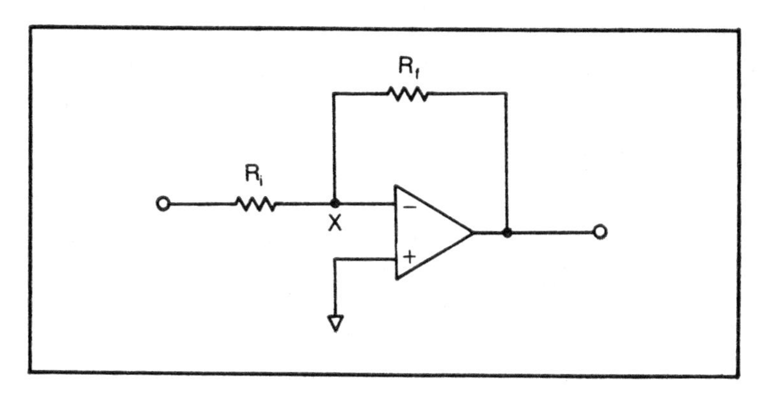

(1) $\frac{R_f \times R_i}{R_i}$ Go to Block 34.

(2) zero volts. Go to Block 18.

BLOCK 13

The correct answer to the question in Block 1 is choice (1). A common-collector amplifier is also called an emitter follower. It always has a voltage gain that is less than 1. It also has a high-input impedance and a low-output impedance.

Here is your next questicn:

The output of a certain push-pull amplifier is supposed to be a pure sine wave. Instead, the output waveform is distorted as shown in this block. This type of distortion is called

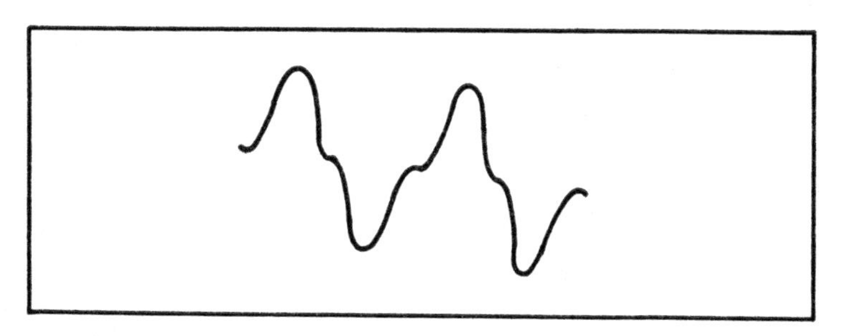

(1) wrinkle finish. Go to Block 39.
(2) crossover distortion. Go to Block 19.

BLOCK 14

Your answer to the question in Block 3 is not correct. Go back and read the question again and select another answer.

BLOCK 15

Your answer to the question in Block 1 is not correct. Go back and read the question again and select another answer.

BLOCK 16

The correct answer to the question in Block 25 is choice (1). Both the tetrode and tunnel diodes have negative resistance regions in their characteristic curves. A diac will break over in two directions, but the four-layer diode will only break over in one direction.

Here is your next question:

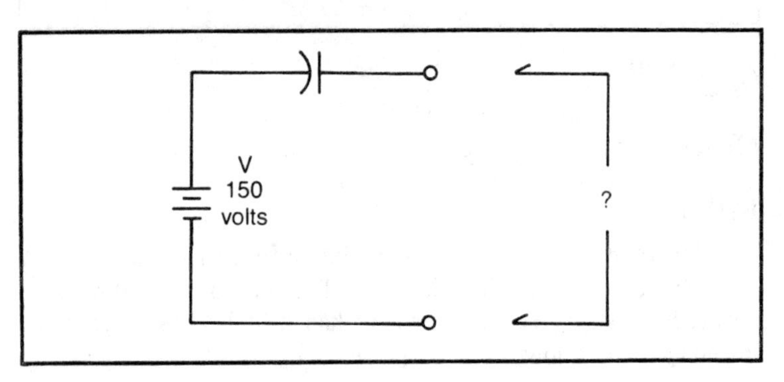

What is the voltage across the output terminals in the circuit shown in this block. (Assume the capacitor is not charged.)

________Go to Block 40.

BLOCK 17

Your answer to the question is Block 38 is not correct. Go back and read the question again and select another answer.

BLOCK 18

The correct answer to the question in Block 12 is choice (2). The point in question is called the summing point. The microprocessor tries to keep that point at a zero-volt level.

Here is your next question:

Analog circuits are also called

(1) linear circuits. Go to Block 38.
(2) digital circuits. Go to Block 9.

BLOCK 19

The correct answer to the question in Block 13 is choice (2). This type of distortion occurs in push-pull amplifiers, and in complementary (totem pole) amplifiers. It is eliminated by applying a small amount of forward bias to the bipolar transistors used in the amplifier. In other words, the power transistors are operated class-AB.

Here is your next question:

The voltage gain (A_v) of the operational amplifier shown in this block is equal to

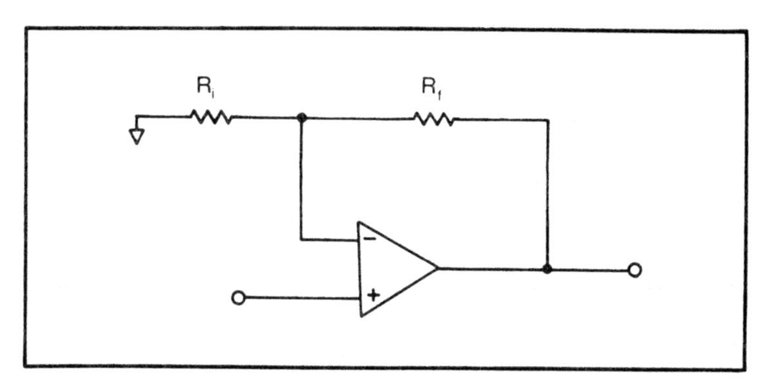

(1) $-\frac{R_f}{R_i}$ Go to Block 29.

(2) $1 + \frac{R_f}{R_i}$ Go to Block 3.

BLOCK 20

Your answer to the question in Block 37 is not correct. Go back and read the question again and select another answer.

BLOCK 21

Your answer to the question in Block 36 is not correct. Go back and read the question again and select another answer.

BLOCK 22

Your answer to the question in Block 25 is not correct. Go back and read the question again and select another answer.

BLOCK 23

Your answer to the question in Block 24 is not correct. Go back and read the question again and select another answer.

BLOCK 24

The correct answer to the question in Block 37 is choice (1). However, only the zener diode operates with a reverse *current.*

Here is your next question:

The circuit in this block is sometimes called a *false bass tone control.* To get more of a bass sound, move the variable resistor (tone control)

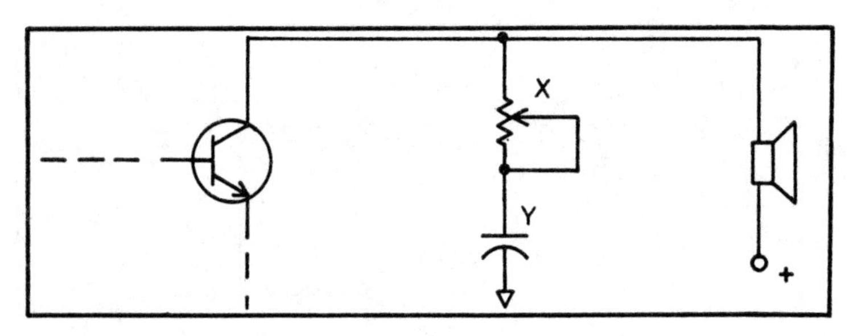

(1) toward point x. Go to Block 11.
(2) toward point y. Go to Block 23.

BLOCK 25

The correct answer to the question in Block 27 is choice (2). That is the definition of rise time.

Here is your next question:

In this question you are only to compare the *shapes* of the characteristic curves. Which of the pairs of characteristic curves would *not* look similar?

(1) Tetrode and tunnel diode. Go to Block 16.
(2) Diac and four-layer diode. Go to Block 22.

BLOCK 26

Your answer to the question in Block 4 is not correct. Go back and read the question again and select another answer.

BLOCK 27

The correct answer to the question in Block 35 is choice (2). NASA proved a long time ago that a mechanical connection is not needed or desired. The only thing you accomplish by wrapping the wire around and through the slot is to make it almost impossible to unsolder the connection without destroying the component.

Here is your next question:

The rise time of a pulse is the time that it takes to go from

(1) minimum to maximum voltage. Go to Block 7.
(2) 10% to 90% of maximum voltage. Go to Block 25.

BLOCK 28

Your answer to the question in Block 6 is not correct. Go back and read the question again and select another answer.

BLOCK 29

Your answer to the question in Block 19 is not correct. Go back and read the question again and select another answer.

BLOCK 30

The correct answer to the question in Block 3 is choice (1). This circuit has an advantage over a full-wave doubler in that it has a zero-volt line that can be connected directly to the power line ground circuit.

Here is your next question:

For the voltage regulator shown here, the voltage at point x will increase (become more positive) when the voltage at point y

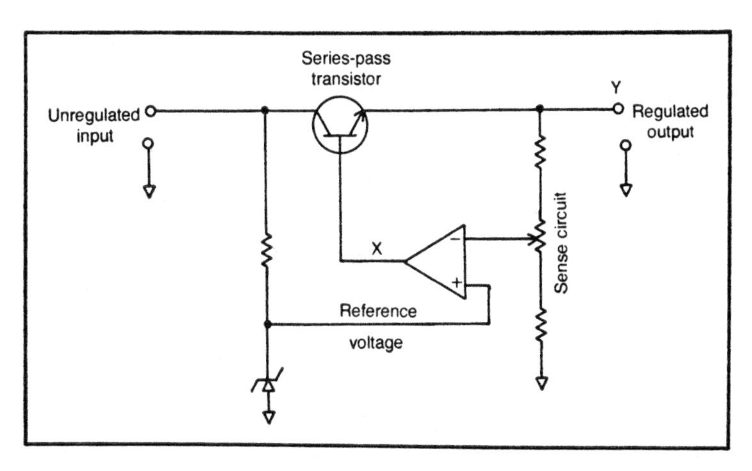

(1) increases. Go to Block 32.
(2) decreases (becomes less positive). Go to Block 12.

BLOCK 31

The correct answer to the question in Block 4 is choice (1). If you missed this question, you should review the specifications for operational amplifiers.

Here is your next question:

Which of the direct-coupled circuits shown in this block will have greater trouble with level shifting?

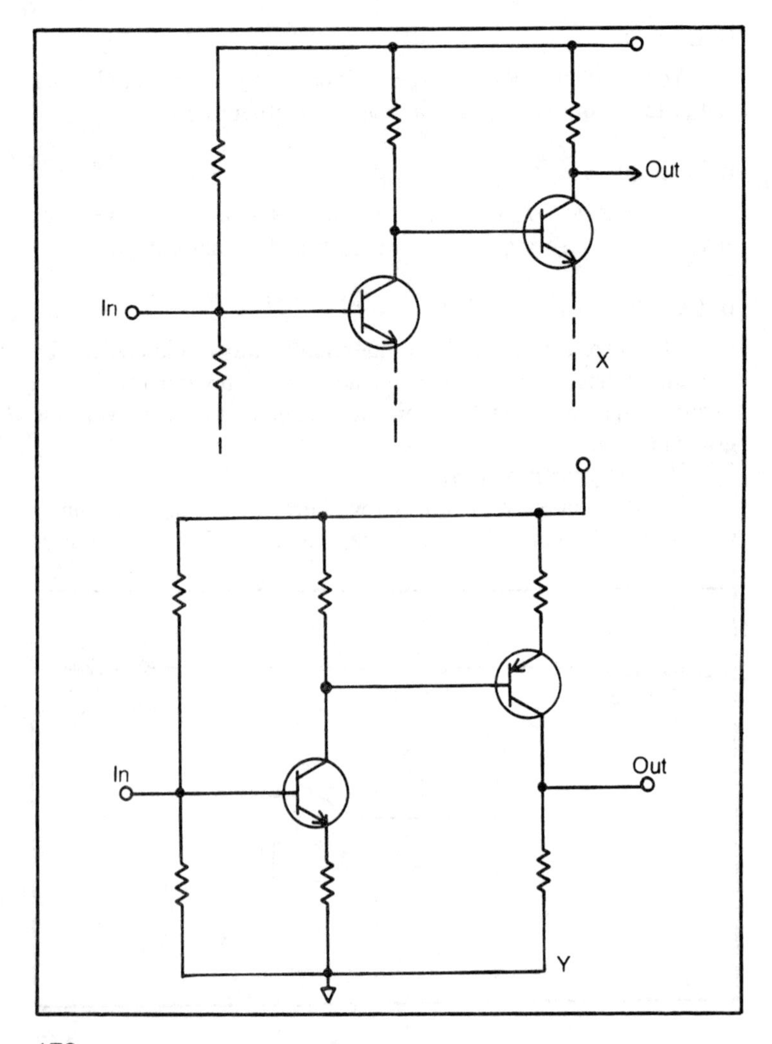

(1) The one marked x. Go to Block 37.
(2) The one marked y. Go to Block 5.

BLOCK 32

Your answer to the question in Block 30 is not correct. Go back and read the question again and select another answer.

BLOCK 33

Your answer to the question in Block 37 is not correct. Go back and read the question again and select another answer.

BLOCK 34

Your answer to the question in Block 12 is not correct. Go back and read the question again and select another answer.

BLOCK 35

The correct answer to the question in Block 6 is choice (2). Of course, there is a limit to the amount of gain you can get this way.

Here is your next question:

The very strong mechanical connection in this illustration is the preferred method prior to soldering. This statement is

(1) correct. Go to Block 8.
(2) not correct. Go to Block 27.

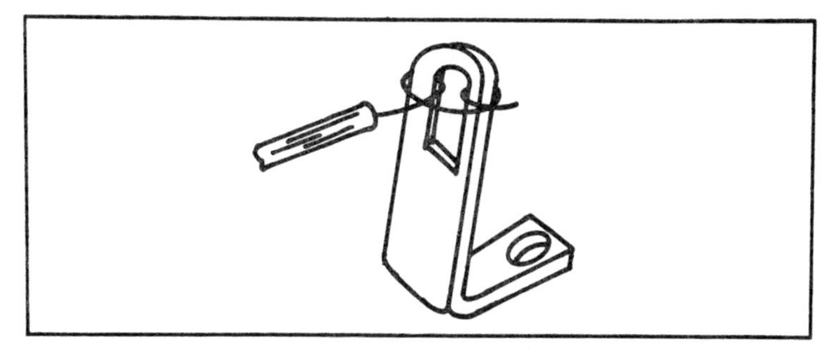

BLOCK 36

The correct answer to the question in Block 11 is choice (2). When the plates are moved closer together, the capacity increases. That increases the value of the denominator in the equation:

$$f_r = \frac{1}{2\pi\sqrt{LC}}$$

Any time you increase the denominator of a fraction, you lower its value (f_r).

Here is your next question:

The circuit in this block is an example of

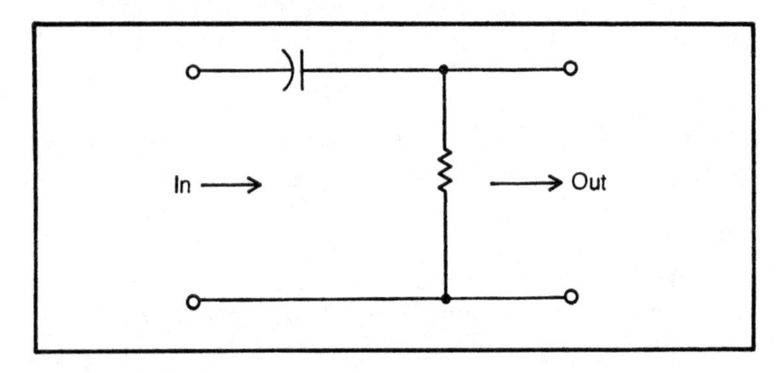

(1) a differentiator. Go to Block 6.
(2) an integrator. Go to Block 21.

BLOCK 37

The correct answer to the question in Block 31 is choice (2). Level shifting is a change in the dc level of direct-coupled amplifiers. In the circuit marked x, the base of the second transistor must be as positive as the collector of the transistor in the previous stage. That will make the collector of the second transistor.

Here is your next question:

Which of the diodes shown in this block is incorrectly voltage biased for its normal operation:

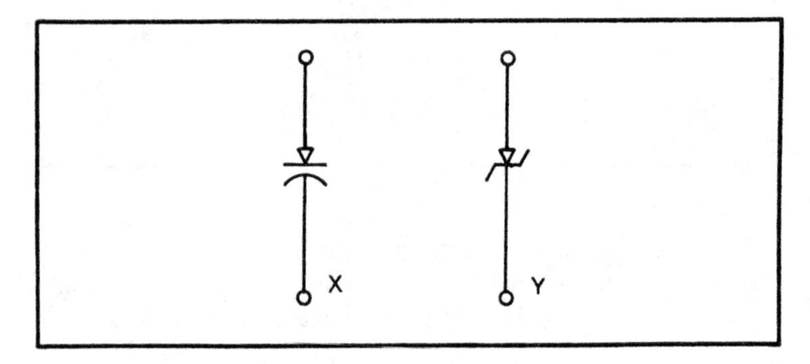

(1) Both are correctly biased. Go to Block 24.
(2) Neither is correctly biased. Go to Block 33.

(3) Only the one shown in x is correctly biased. Go to Block 20.
(4) Only the one shown in y is correctly biased. Go to Block 2.

BLOCK 38

The correct answer to the question in Block 18 is choice (1). The word linear is used more.

Here is your next question:

In order to get the two transistors of a differential amplifier to produce identical voltages at the output terminals, you may have to adjust the

(1) input bias. Go to Block 4.
(2) slewing rate. Go to Block 17.

BLOCK 39

Your answer to the question in Block 13 is not correct. Go back and read the question again and select another answer.

BLOCK 40

The correct answer to the question in Block 16 is 150 volts! This has been the subject of very much discussion among members of ISCET. However, the answer is correct as given here.

One way to understand why the answer is correct is to go around the loop as with Kirchhoff's voltage law. The algebraic sum of the voltages around any closed loop is zero. If there is no voltage drop across the (uncharged) capacitor, then the voltage drop across the output terminals must equal the voltage rise of the battery.

You have now completed the programmed section.

ADDITIONAL PRACTICE CET TEST QUESTIONS

01. Feedback in a blocking oscillator is through a

(1) coupling capacitor.
(2) diode.
(3) transformer
(4) saturated transistor or tube.
(5) a coaxial cable.

02. When the base of a transistor is the same voltage as the emitter, the transistor is

(1) saturated.
(2) operating normally.
(3) cut off.

03. The free-running frequency of a synchronized oscillator should be

(1) exactly equal to the synchronizing frequency.
(2) slightly below the synchronizing frequency.
(3) slightly above the synchronizing frequency.

04. The greatest portion of a superheterodyne receiver's gain is accomplished in

(1) the rf amplifier.
(2) the i-f amplifier.
(3) the detector stage.
(4) the video amplifier.
(5) the voltage amplifier.

05. Which of the following fm detector circuits is insensitive to amplitude modulation?

(1) A Foster-Seeley fm discriminator
(2) A crystal-diode shunt-type video detector
(3) A ratio detector
(4) A slope detector
(5) A homodyne

06. A circuit in which a single amplifier acts as both a sound i-f amplifier and an audio amplifier is called

(1) an impossibility.
(2) a reflex-amplifier circuit.
(3) a reflectodyne circuit.
(4) a neutrode circuit.
(5) an amplidyne.

07. Radiation of the local oscillator signal from a superheterodyne receiver is prevented (or greatly reduced) by the use of

(1) reducing the dc voltage to a local oscillator.
(2) twisting the twin-lead line from the antenna.
(3) an rf amplifier.
(4) forward agc.
(5) a shielded antenna.

08. An advantage of using a pentode over a triode for an rf amplifier is that

(1) it does not require neutralization.
(2) it is more linear.
(3) it produces a lower amount of noise.

(4) it will operate with a lower power.
(5) it is much cheaper.

09. The i-f frequency of an fm broadcast receiver is

(1) 41.25 MHz.
(2) 10.7 MHz.
(3) 3.58 kHz.
(4) 455 kHz.
(5) 4.5 kHz.

10. Which of the following is the i-f frequency for a standard am broadcast receiver?

(1) 10.7 MHz
(2) 10.7 kHz
(3) 540 kHz
(4) 0.455 MHz
(5) None of these answers is correct.

11. When the television primary colors are combined in the color tube so that $E_y = 0.59_G + 0.30E_R + 0.11_B$ the result is

(1) a white screen.
(2) a green color on the screen.
(3) a black screen.

12. Another name for a Darlington connection is

(1) alpha squared.
(2) beta squared.
(3) gamma squared.
(4) delta squared.
(5) omega squared.

13. An integrator is a

(1) circuit that combined two different kinds of oscillator operations.
(2) low-pass filter.
(3) circuit that combines two different kinds of pulses.
(4) type of delay line.
(5) high-pass filter.

14. A ferrite bead behaves like

(1) an inductor.
(2) a temperature-sensing resistor.

15. The filter circuit shown in Fig. 6-13 is a

(1) high-pass type.
(2) low-pass type.

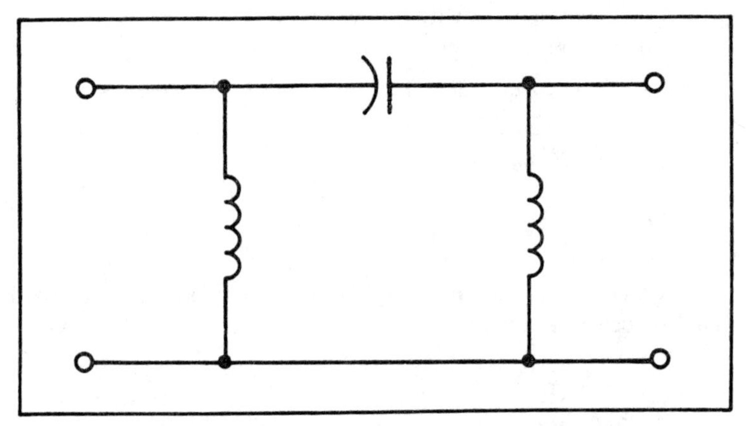

Fig. 6-13. You should be able to identify filter circuits like this one before you take the CET test.

16. For the amplifier shown in Fig. 6-14 the output signal is

(1) in phase with the input signal.
(2) 180° out of phase with the input signal.

17. Which amplifier configuration will introduce a 180° phase shift to the signal?

(1) Common drain
(2) Cathode follower
(3) Grounded grid

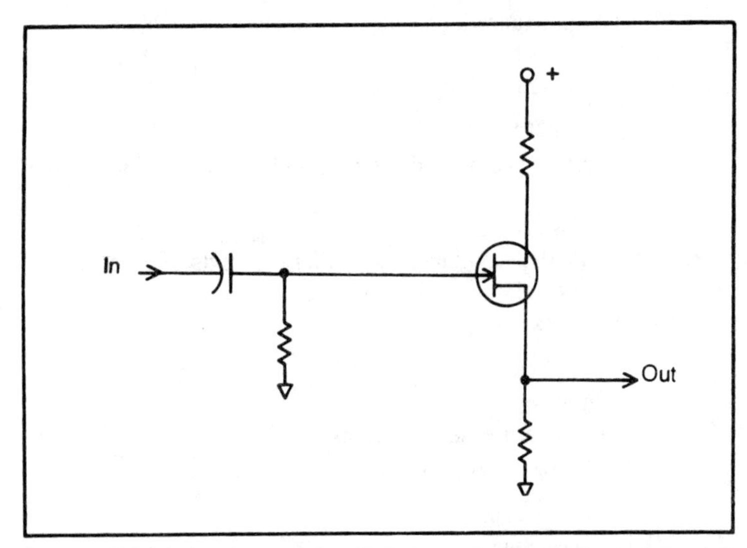

Fig. 6-14. What is the phase relationship between the input and output signals?

(4) Common source
(5) Common base

18. Which amplifier configuration can be used to match a high impedance to a low impedance?

(1) Grounded base
(2) Conventional
(3) Emitter follower
(4) Common source
(5) Grounded gate

19. You would expect to obtain the best high-frequency response from a

(1) conventional amplifier.
(2) grounded-grid amplifier.
(3) grounded-emitter amplifier.
(4) grounded-source amplifier.
(5) grounded drain.

20. Class-A operation is not possible with

(1) contact-biased tube circuits.
(2) grid-leak-biased tube circuits.

21. Which tube type would normally produce the greatest amount of noise in an amplifier circuit?

(1) Pentagrid
(2) Pentode
(3) Tetrode
(4) Triode

22. Which type of bias could result in a destroyed tube if the input signal is lost?

(1) Contact bias
(2) Automatic bias
(3) Grid-leak bias
(4) C-battery bias

23. Removing the capacitor across an emitter resistor in a transistor amplifier will

(1) increase the gain of the amplifier.
(2) increase the amplifier frequency response.

24. The emitter response in a bipolar transistor circuit is used for

(1) reducing amplifier noise.
(2) increasing the gain of the stage.

(3) biasing the transistor.
(4) stabilizing the stage against the effects of temperature change and against thermal runaway.
(5) positive feedback.

25. A bootstrap circuit for an amplifier input circuit is used to

(1) decrease the frequency response of the amplifier.
(2) increase the input impedance of the amplifier.
(3) decrease the output impedance of the amplifier.
(4) increase the output impedance of the amplifier.
(5) decrease the input impedance of the amplifier.

26. Peaking compensation is used to

(1) increase bandwidth.
(2) improve fidelity of audio amplifiers.
(3) reduce noise.
(4) increase the high-frequency gain of an amplifier.
(5) None of these choices is correct.

ANSWERS TO PRACTICE TEST

Question Number	Answer Number
01	(3)
02	(3)
03	(2)
04	(2)
05	(3)
06	(2)
07	(3)
08	(1)
09	(2)
10	(4)
11	(1)
12	(2)
13	(2)
14	(1)
15	(1)
16	(1)
17	(4)
18	(3)
19	(2)
20	(2)
21	(1)
22	(3)
23	(2)
24	(4)
25	(2)
26	(4)

Television

Experienced journeyman television technicians do not normally have trouble answering questions in this section of the CET test. Monochrome and color television circuitry are now included together in the television section. You may be asked questions about the television signal and that information is also included in this review.

A separate section in the CET test has five questions on videocassette recording and videodisc recording. Material for that section is also reviewed in this chapter.

TELEVISION AND FM SIGNALS

The distribution of frequencies in a television channel are shown in Fig. 7-1. Note that there is a 1.25 megahertz vestigial sideband transmitted as part of this signal. This vestigial sideband and the carrier are included with the transmitted signal to simplify tuning of the television receiver. *Neither carries any useful information as far as reproducing the picture or sound is concerned.*

The 4.5 megahertz frequency range between the picture carrier and the sound carrier is a very important thing to remember. In a monochrome receiver it is not uncommon to use a product detector to separate the sound signal. In that circuit, the picture carrier is heterodyned with the sound signal to produce a sound i-f signal. This is not done in a color TV receiver. The reason is that *the sound must be taken off at some point before the detector in order to avoid*

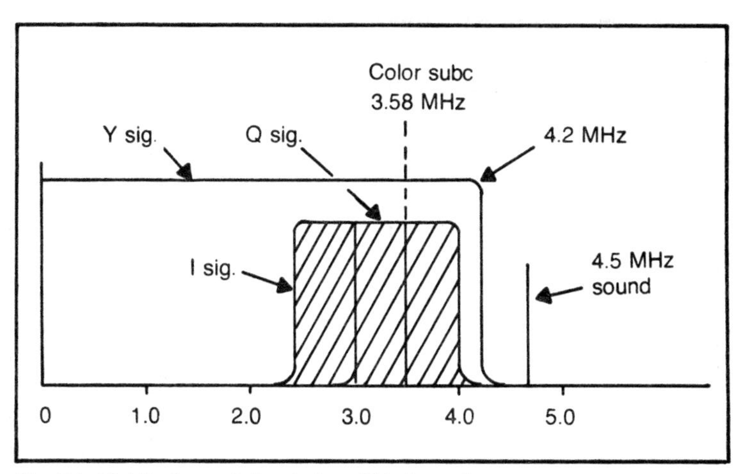

Fig. 7-1. Distributions of signals for a TV channel.

heterodyning between the sound and 3.58 megahertz chroma signals.

Figure 7-2 shows the distribution of television signals after they pass through the tuner. Note that the video carrier (at 45.75 megahertz) is now *above* the sound carrier in frequency. This is opposite to condition shown in Fig. 7-1.

To understand why the positions are reversed, refer to Fig. 7-3. In this illustration the video and sound frequencies for channel 33 are shown. The local oscillator frequency that is required to produce a 45.75 megahertz video i-f frequency is also shown. Note that the difference between the local oscillator and sound signal is less than the difference between the local oscillator and the video signal.

Another way of saying this is that the sound signal will have a lower frequency than the video frequency after the signal has been heterodyned with the local oscillator in the mixer stage. This is always true. It is also the general case that the local oscillator frequency is *above* the sound and video signal.

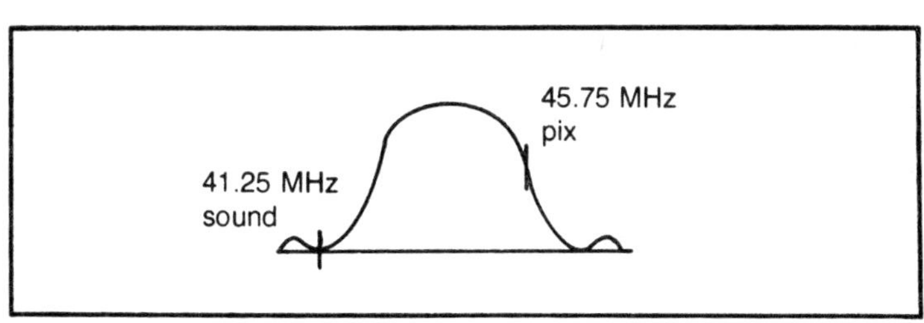

Fig. 7-2. After the signal passes through the tuner the positions of the sound and video carriers are reversed.

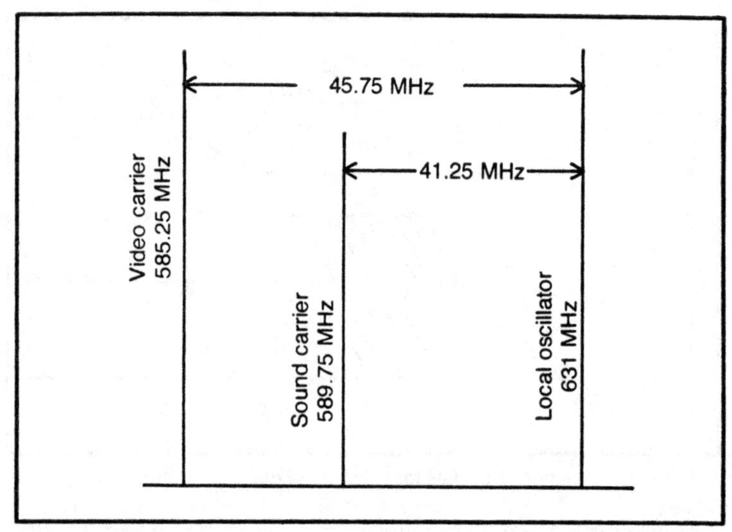

Fig. 7-3. The reason for the carrier position reversal is illustrated here. The frequencies are for channel 33.

Figure 7-4 shows the distribution of signal strength for video blanking pedestal and synchronizing pulses. Observe from this illustration that the signal is never allowed to go below 12.5 percent of maximum. Since it is an amplitude modulation signal this can be stated another way—*100 percent modulation is never achieved with the television signal.*

The 8 cycles of color burst are located on the back porch of the horizontal synchronizing and blanking pedestal. A minimum of 8 cycles is transmitted for this burst. In reality, this burst signal is the

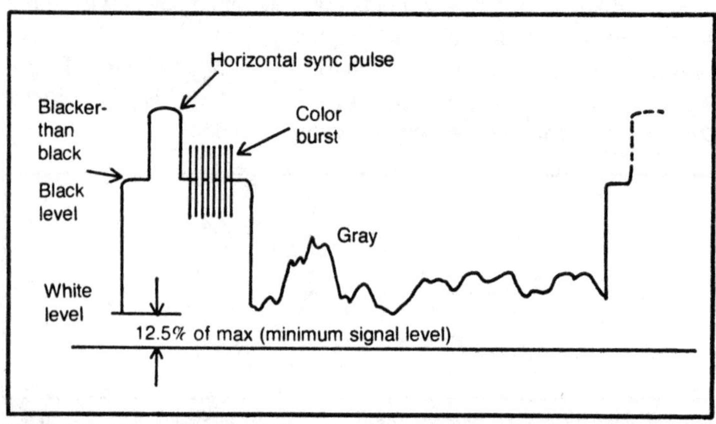

Fig. 7-4. Relative amplitudes of the video and sync pulse signals are shown here.

original color carrier of the transmitter system. As you know, the color carrier is suppressed and only sideband information is sent for the color signals. Therefore, some form of *carrier reinsertion* is necessary. The color burst offers the reference that is necessary for the color carrier to be reinserted in the proper frequency and phase relationship that it has had at the transmitter.

Two important signals are sent during the blanking period between television fields. You will remember that two fields required to make one *frame,* or complete picture. Each field is actually a half picture obtained by scanning the odd lines for one field and the even lines for the next field. The blanking period is transmitted as a blanking pedestal which shows up as a dark space if you roll the picture vertically to a half of a frame.

The *VITS* signal (vertical interval test signal) is used primarily for evaluating the quality of television signals. Experienced journeyman CET technicians have indicated to me during a lecture tour in 1983 that they are using this VITS signal as one method of evaluating receiver performance.

A second signal that is sent during the blanking period is the *VIRS* (vertical interval reference signal). *This signal is used in some receivers to automatically adjust the color of the receiver so that it closely matches the color at the transmitter.* Of course, special VIRS signal processing is required for this feature.

In a monochrome system the horizontal sweep frequency is generally given at 15,750 hertz. That frequency had to be modified slightly when the color signal was added into the NTSC monochrome signal. However, the 15,750 is still generally given as the horizontal oscillator frequency.

Using the equation:

$$T = \frac{1}{f}$$

the time for the horizontal sweep is then found to be

$$T = \frac{1}{15{,}750} = 63.4 \text{ microseconds}$$

That is the time between horizontal sync pulses. It is often designated H and used as a time reference for all TV signals. It includes time for the blanking pedestal, sync pulse, and one line of video.

There are 60 fields sent per second so the time for one field is also easily computed by the equation

$$T = \frac{1}{f}$$

using a field frequency of 60 hertz the time for one field becomes

$$T = \frac{1}{60} = 16.667 \text{ milliseconds}$$

It takes two fields to make one frame. Therefore, the time for one complete frame is equal to twice the time for one field: Time for Frame = 2 × 16.667 ms = 33.333 milliseconds

COLOR TV

The color television picture is made up of three color primaries: red, green, and blue. These are *additive* primaries. They are not to be confused with the type of primary colors that you used when you were playing with crayons. Those were *subtractive* primaries. With subtractive primaries the color green and blue combine to make yellow. But, with additive primaries the colors green and blue combine to make the color cyan.

Of special importance is *Illuminant C*. That is a term that refers to the equavalent of daylight. To understand why this is important consider, what would happen if you added the red, green and blue colors in equal amounts. The result would be a white that has a very heavy blue tint, and it would be uncomfortable to watch.

When you are watching a color picture of a daylight scene you would like the sunlit area to match as closely as possible the sunlit areas in your real life experience. To accomplish this, the three primary colors must be added in the following proportions: Illuminant C = 0.11 Blue + 0.59 Green + 0.30 Red.

FM

As a consumer electronics technician you should at least be familiar with the makeup of the FM multiplex signal. This is shown in Fig. 7-5

The *pilot* in this signal is used only as reference so that the 38 kilohertz signal carrier can be reinserted. In the receiver, this pilot signal is doubled in frequency and used for carrier reinsertion to demodulate the L-R signal.

The 67 kilohertz SCA subcarrier is used for transmitting music without accompanying advertisements. An SCA decoder would be used for playing music in places like a physician's waiting room.

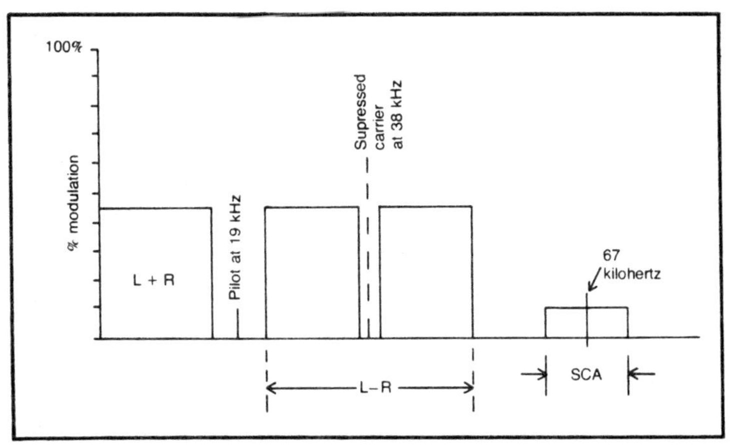

Fig. 7-5. The distribution of signals in an FM stereo channel shows how L+R and L−R signals are transmitted.

THE TELEVISION RECEIVER

As a consumer electronics technician you should be familiar with the monochrome and color receivers in block diagram form. Figure 7-6 shows the block diagram of a monochrome receiver. In the CET test you might be shown a block diagram like this and be asked to identify the names of blocks in various parts of the receiver. Figure 7-7 is a highly generalized color TV receiver that does not represent any particular brand-name. As with the monochrome block diagram, you may be asked to identify blocks in this type of drawing.

The incoming signal is delivered to the UHF and VHF tuner where it is mixed with the local oscillator signal in a stage called the *mixer* to produce the intermediate frequency signal. This intermediate frequency signal is amplified in several stages. Just before the video detector the sound signal is separated off to be processed in a separate section of the receiver.

In the particular block diagram the output of the video detector goes to three different places. One is to the delay line luminance amplifier. This is simply the video amplifier with a different name.

The second output from the video detector goes to the color *bandpass amplifier*. Its purpose is to eliminate most of the luminance signal and pass only the I and Q color signals. After these signals have been amplified they are decoded and delivered to a matrix (decoded) for distribution to the three color grids in the color picture tube. In some receivers the luminant signal had been combined in a matrix to produce the red, green, and blue signals. In

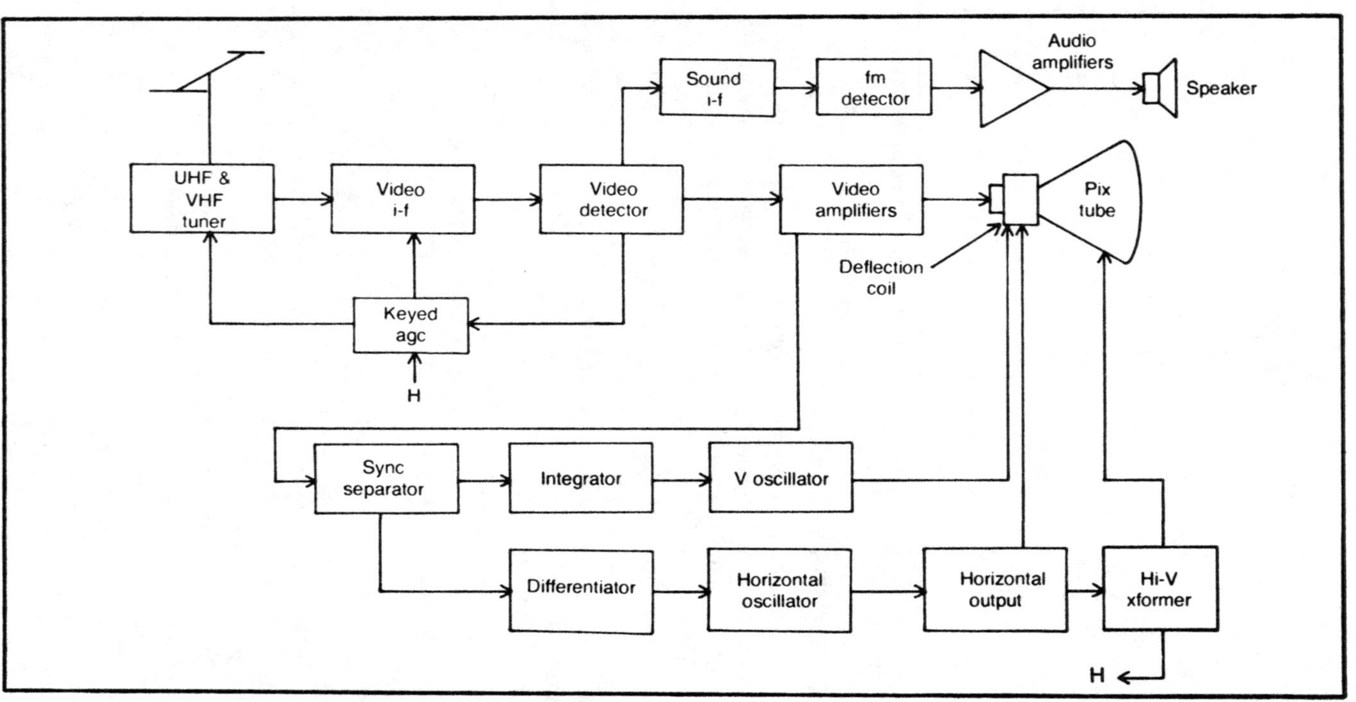

Fig. 7-6. This is a simplified block diagram of a monochrome receiver.

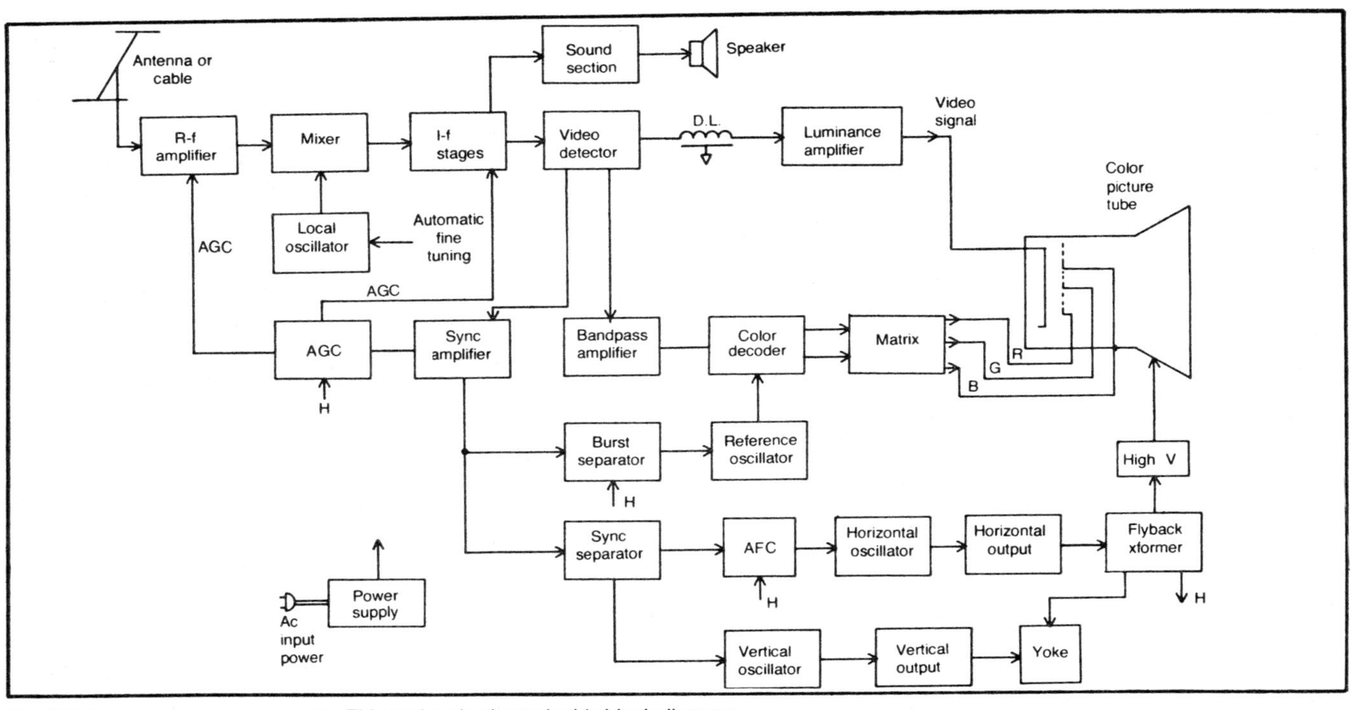

Fig. 7-7. A general version of a color TV receiver is shown in this block diagram.

another type of receiver these signals are combined inside of the television picture tube.

Another output of the detector goes to the *sync amplifier.* It produces synchronizing signals for the sync separator and also for the burst separator. Remember that the burst sits on the back porch of the horizontal blanking pedestal. It is removed by a coincidence circuit that has an input pulse from the horizontal flyback transformer. When the horizontal pulse arrives it permits the burst amplifier to amplify the burst on the back porch. At all other times during the sync signal the H input is not present. Therefore, there is no output from this amplifier at all other times.

The burst signal is used to synchronize the reference oscillator then the oscillator signal is recombined with the I and Q quadrature signals to produce signals which are similar to the ones in the transmitter before the carrier was removed.

The sync separator produces an output signal in both the horizontal and vertical oscillator sections. A differentiating circuit is usually used to remove the horizontal sync signals. An integrating circuit is used to separate out the *vertical* sync signals.

In the horizontal section there is an automatic frequency control (afc) which holds the horizontal oscillator to the correct frequency even though noise may be present in this section. The output of the horizontal oscillator goes to a horizontal output stage and then to the *flyback transformer.*

The flyback transformer produces the very high voltage necessary for the dc second anode voltage on the picture tube. It is also the source of the signal that is used for various processing in the color receiver.

The vertical oscillator and vertical output signal, as well as the horizontal signals, are delivered to the yoke. It moves the beam back and forth and up and down on the screen. The rectangular area on the screen that is swept by the deflected beam is called the raster.

Although not shown in the illustration an automatic frequency control circuit may be present in the tuner. It gets its signal from the i-f stage and is used to lock the local oscillator signal onto the frequency required to demodulate the carrier.

Ideally the local oscillator is always an exact frequency above the carrier frequency of the incoming signal. If the oscillator drifts, the color picture portion of the signal can very quickly be downgraded or lost. So, this afc circuit (also called aft for automatic fine tuning) is designed primarily to maintain color television reception.

The automatic gain control (agc) in the receiver is *keyed.* That means the strength of the agc signal depends upon the strength of the sync pulses on the incoming signal. It is difficult, if not impossible, to get an agc voltage off of the video signal because the picture changes continuously with the brightness and darkness of the scenes change. Using the sync pulses as a reference is more logical because their amplitude is directly related to signal strength. That is what is done on the keyed agc circuit.

The horizontal pulse (H) from the flyback transformer turns the agc circuit on only during the time when the horizontal sync pulses are present. In other words, the agc amplifier is keyed on by the horizontal signal. Therefore the output consists only of the sync pulses. The sync pulses are rectified to produce a dc voltage that controls the gain of the rf amplifier and i-f amplifiers in the system.

The power supply in the block diagram converts the ac line power into dc voltages that are necessary for operating the various sections of the receiver. A disadvantage of this type of system is that the power transformer is very heavy and very costly.

A more modern system uses what is known as a *scan-derived power supply.* In that system the dc voltages for operating the various stages is obtained by rectifying the ac voltage pulses from the high voltage transformer. With that type of circuit it is necessary to have a startup circuit. In other words, you won't get any dc out of the flyback transformer until the horizontal oscillator and horizontal output stages are working.

So, a starting circuit is necessary to produce a dc voltage to those circuits to get them into operation. Once the dc voltage is obtained from the flyback the startup circuit is automatically disconnected and the circuit operates primarily from the scan-derived voltages obtained from the output transformer.

Figure 7-8 shows a block diagram of somewhat different color television receiver design. The tuners and the i-f module are virtu ally the same. Likewise the sound module which receives its signal from the i-f stage before the detector is also similar to the one previously discussed.

The signal is detected and the output video signal is delivered to a comb filter module. The comb filter is primarily designed to separate the color signal from the luminance signal. As you can see in Fig. 7-1 these signals exist in the same region for the spectrum of

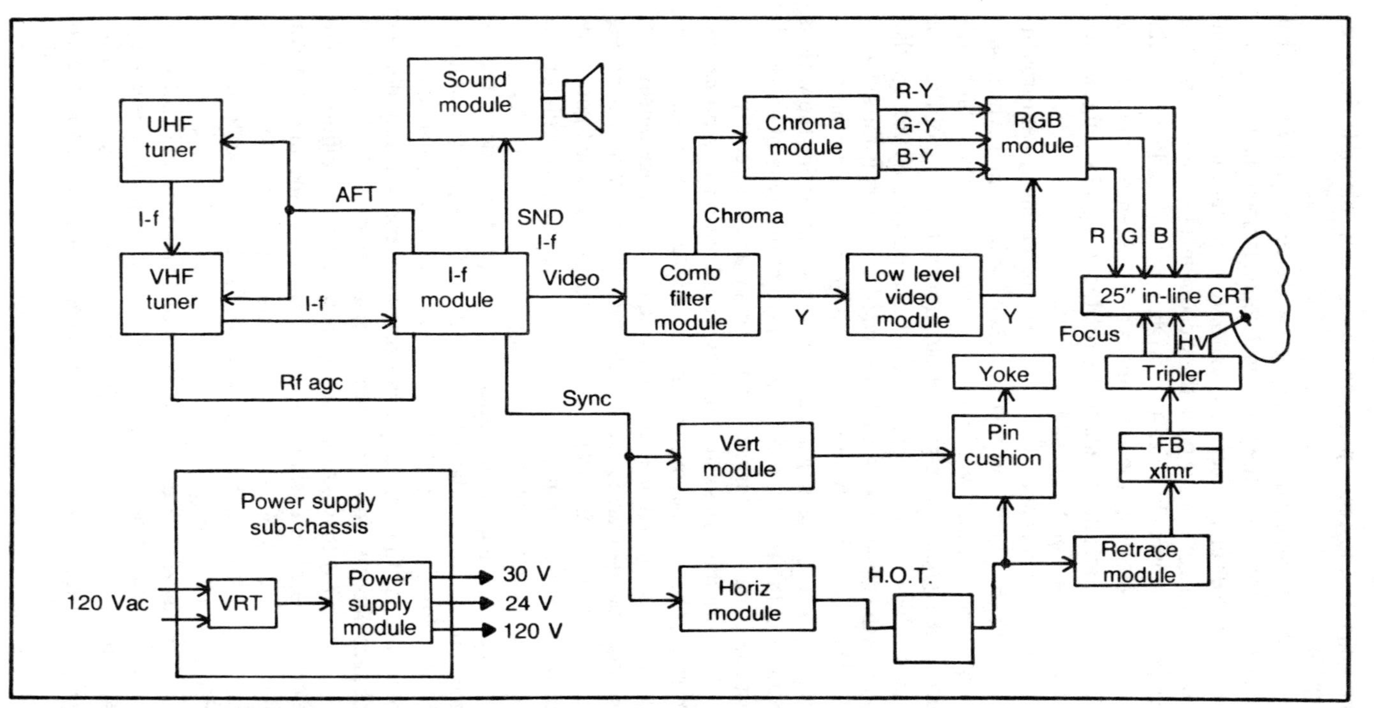

Fig. 7-8. A comb filter may be used to separate the color signals from the composite video.

the television bandwidth. Special circuitry must be used to keep each of these signals separated in the receiver.

One way to do this is to use filters in the bandpass amplifier to keep out the luminance signal. Likewise, filters can be used in the luminance section to keep out the color signal. The comb filter does this much more efficiently. Essentially, it is a very sharp filter that eliminates interference between those two portions of the signal.

Note that there are two outputs from the comb filter. One goes to the *chroma module* where the color signal is demodulated to produce the red, green, and blue signal voltages. The other produces the Y, or luminance signal. That signal also goes to the matrix. In this particular receiver the luminance and the color signals are combined in the RGB module and then delivered to the picture tube.

The rest of this color receiver is similar to the one discussed with reference to the previous block diagram. Note that pincushion correction is obtained in a pincushion module. This module produces signals that prevent sagging in the vertical and horizontal perimeter of the raster.

Not shown in the block diagrams is the color killer circuit. It's purpose is to shut off the color amplifiers when there is no color signal being received. The reason this is important is that it eliminates the so-called *colored confetti* (which is colored snow) produced by amplifiers in the color section. This colored confetti is generated by the bandpass amplifier in the absence of an incoming color signal.

Remember, the color killer is OFF when the bandpass amplifier is passing a signal. The color killer comes ON to shut the amplifier OFF in the absence of a color signal.

VIDEO RECORDERS AND THE CET TEST

There is a 5-question section in the consumer test regarding video recorders. This is primarily a section on basic fundamentals and does not include any advanced troubleshooting techniques. Nor is there an extensive amount of circuitry involved at this time.

You should understand that the video signal is *recorded as an FM signal* in video tape recording systems. You should also understand that in the tape system the *sound is recorded on a separate track.*

A helical scanning system is used with rotating heads to reduce the necessity for very high speed tape transport. In other words the tape is moved in one direction the head is scanned across the tape in the opposite direction. This arrangement provides a very high tape-to-head relative speed.

A capstan type of drive system is used, as in audio tape systems, to make the tape speed independent of the amount of tape that's on the reel. If it wasn't for the capstan drive, the tape speed would change as the diameter of the tape on one reel decreased and, and at the same time, increased on the other reel.

In the video disc system a laser beam may be used to sense the information on one disc. It may also be used to maintain the pickup section in the center of the groove.

If you encounter specialized circuits in this part of the test such as those used to monitor tape slack and tape transport, you will be able to analyze these circuits on the basis of fundamentals and no special knowledge of tape systems will be required. At the present time there is a proposal for a separate CET Test with a specialty in Video Recording.

PROGRAMMED REVIEW

Start with block number 1. Pick the answer that you feel is correct. If you select choice number 1, go to block 13. If you select choice number 2, go to block 15. Proceed as directed. There is only one correct answer for each question.

BLOCK 1

The choice of an intermediate frequency for a television receiver is based upon

(1) image frequency and bandwidth considerations. Go to Block 13.
(2) whether tubes or transistors are used in i-f amplifier stages. Go to Block 15.

BLOCK 2

The correct answer to the question in Block 16 is choice (2). Remember that the *saturation* of a color is determined by the *amplitude* of the signal, and the hue (actual color) is controlled by the phase of the subcarrier with reference to the I and Q signals.

Here is your next question:

Which of the following signals may be used to automatically adjust the color in a receiver?

(1) VITS. Go to Block 8.
(2) VIRS. Go to Block 18.

BLOCK 3

Your answer to the question in Block 37 is not correct. Go back

and read the question again and select another answer.

BLOCK 4

Your answer to the question in Block 13 is not correct. Go back and read the question again and select another answer.

BLOCK 5

Your answer to the question in Block 30 is not correct. Go back and read the question again and select another answer.

BLOCK 6

The correct answer to the question in Block 27 is choice (1). The AGC control is used for r-f and i-f amplifiers.

Here is your next question:

Which of the following is used to prevent ringing in the flyback transformer?

(1) an LDR. Go to Block 9.
(2) a damper diode. Go to Block 17.

BLOCK 7

The correct answer to the question in Block 21 is choice (2). Hot carrier diodes do not have anything to do with temperature. They are made with a metal/semiconductor interface, and they have a very low forward voltage drop.

Here is your next question:

If the R-C differentiator is not working properly there will be a loss of

(1) horizontal sweep. Go to Block 33.
(2) vertical sweep. Go to Block 40.

BLOCK 8

Your answer to the question in Block 2 is not correct. Go back and read the question again and select another answer.

BLOCK 9

Your answer to the question in Block 6 is not correct. Go back and read the question again and select another answer.

BLOCK 10

The correct answer to the question in Block 13 is choice (2).

There is no resultant phase inversion of the signal from the detector to the cathode. A positive-going signal on the cathode of a picture decreases the brightness on the screen.

Here is your next question:

In the automatic fine tuning circuit (aft) you are likely to find

(1) a varactor diode. Go to Block 16.
(2) an SCR. Go to Block 26.

BLOCK 11

Your answer to the question in Block 37 is not correct. Go back and read the question again and select another answer.

BLOCK 12

Your answer to the question in Block 22 is not correct. Go back and read the question again and select another answer.

BLOCK 13

The correct answer to the question in Block 1 is choice (1). Lower intermediate frequencies result in more of a problem getting the required bandwidth, and also cause greater problems with image frequencies. However, a higher receiver gain is possible with a lower i-f frequency.

Here is your next question:

In a certain television receiver the video signal is delivered to the cathode of the picture tube; and, there are two common emitter video-amplifier stages along with one emitter follower. The output signal from the video detector should have

(1) the sync pulses as the most negative part of the signal. Go to Block 4.
(2) the sync pulses as the most positive part of the signal. Go to Block 10.

BLOCK 14

Your answer to the question in Block 21 is not correct. Go back and read the question again and select another answer.

BLOCK 15

Your answer to the question in Block 1 is not correct. Go back and read the question again and select another answer.

The correct answer to the question in Block 10 is choice (1). The varactor diode is sometimes known by the trade name Varicap. It controls the local oscillator frequency.

Here is your next question:

The tint control in a color receiver

(1) adjusts the gain of the bandpass amplifier. Go to Block 35.
(2) adjusts the phase of the color subcarrier. Go to Block 2.

BLOCK 17

The correct answer to the question in Block 6 is choice (2). An LDR is a Light-Dependent Resistor. It is not used in the flyback circuit, but you might find one in an automatic brightness control circuit.

Here is your next question:

Where would you be more likely to find a noise inverter?

(1) in the video amplifier. Go to Block 23.
(2) in the horizontal sweep section. Go to Block 28.

BLOCK 18

The correct answer to the question in Block 2 is choice (2). The Vertical Internal Reference Signal contains information on the color reference.

Here is your next question:

Why is a keying pulse delivered to the color bandpass amplifier?

(1) to decode the signal. Go to Block 24.
(2) to remove the burst. Go to Block 29.

BLOCK 19

Your answer to the question in Block 33 is not correct. Go back and read the question again and select another answer.

BLOCK 20

Your answer to the question in Block 29 is not correct. Go back and read the question again and select another answer.

BLOCK 21

The correct answer to the question in Block 39 is choice (2).

Here is your next question:

To prevent temperature changes from affecting yoke performance, you will find the following component in the yoke circuitry:

(1) a hot-carrier diode. Go to Block 14.
(2) a thermistor. Go to Block 7.

BLOCK 22

The correct answer to the question in Block 31 is choice (2). This question would have to be changed for the CET test because not every receiver has a vertical switch transistor.

Here is your next question:

Normally, feedback in a Blocking Oscillator is accomplished with

(1) a transformer. Go to Block 30.
(2) an R-C circuit. Go to Block 12.

BLOCK 23

Your answer to the question in Block 17 is not correct. Go back and read the question again and select another answer.

BLOCK 24

Your answer to the question in Block 18 is not correct. Go back and read the question again and select another answer.

BLOCK 25

Your answer to the question in Block 33 is not correct. Go back and read the question again and select another answer.

BLOCK 26

Your answer to the question in Block 10 is not correct. Go back and read the question again and select another answer.

BLOCK 27

The correct answer to the question in Block 37 is choice (1). This signal is also known as the video or video signal.

Here is your next question:

A certain receiver has seven direct-coupled video amplifiers. You would *not* expect to find the following control in the video amplifier stage.

(1) the AGC bias control. Go to Block 36.
(2) the brightness control. Go to Block 6.

BLOCK 28

The correct answer to the question in Block 17 is choice (2). Noise pulses can accidentally trigger the horizontal oscillator, so it is a good idea to get rid of them.

Here is your next question:

Which of the following is the more likely source of a keying pulse in the AGC circuit?

(1) the flyback transformer. Go to Block 31.
(2) the video detector. Go to Block 38.

BLOCK 29

The correct answer to the question in Block 18 is choice (2).

Here is your next question:

The I and Q signals have the same frequency and they are

(1) 90° out of phase. Go to Block 37.
(2) 180° out of phase. Go to Block 20.

BLOCK 30

The correct answer to the question in Block 22 is choice (1). There are a number of similarities between a Blocking oscillator and an Armstrong oscillator.

Here is your next question:

Interlacing in a television receiver is controlled by the

(1) keying pulses. Go to Block 5.
(2) equalizing pulses. Go to Block 39.

BLOCK 31

The correct answer to the question in Block 28 is choice (1). The keying pulse is needed to separate the sync pulses from the composite video signal.

Here is your next question:

In the vertical sweep circuit the vertical switch transistor

(1) starts the sweep oscillator for each new field. Go to Block 34.
(2) discharges the integrator capacitor. Go to Block 22.

BLOCK 32

Your answer to the question in Block 39 is not correct. Go back and read the question again and select another answer.

BLOCK 33

The correct answer to the question in Block 7 is choice (1). The

RC circuit associated with the vertical sweep is called an integrator.

Here is your next question:

Which of the following is a type of relaxation oscillator?

(1) Clapp. Go to Block 25.
(2) Synchrotron. Go to Block 19.
(3) UJT. Go to Block 41.

BLOCK 34

Your answer to the question in Block 31 is not correct. Go back and read the question again and select another answer.

BLOCK 35

Your answer to the question in Block 16 is not correct. Go back and read the question again and select another answer.

BLOCK 36

Your answer to the question in Block 27 is not correct. Go back and read the question again and select another answer.

BLOCK 37

The correct answer to the question in Block 29 is choice (1).

Here is your next question:

Which of the following is another name for the luminance signal?

(1) the Y signal. Go to Block 27.
(2) the AGC output. Go to Block 11.
(3) the Q signal. Go to Block 3.

BLOCK 38

Your answer to the question in Block 28 is not correct. Go back and read the question again and select another answer.

BLOCK 39

The correct answer to the question in Block 30 is choice (2). The *transmitted* equalizing pulses are used to assure interlacing.

Here is your next question:

The television picture will roll down if the vertical oscillator frequency is too

(1) low. Go to Block 32.
(2) high. Go to Block 21.

BLOCK 40

Your answer to the question in Block 7 is not correct. Go back and read the question again and select another answer.

BLOCK 41

The correct answer to the question in Block 33 is choice (3). A relaxation oscillator is made with an R-C or R-L time constant circuit. Also, it has a non-sinusoidal output waveform.

Here is your next question:

Will a sawtooth voltage produce a sawtooth circuit in the horizontal yoke?

____________yes or no. Go to Block 42.

BLOCK 42

The correct answer to the question in Block 41 is *NO*. To get a sawtooth current in the yoke it is necessary to use a waveform that has a sawtooth voltage sitting on top a rectangular wave. This special waveform is called a *trapezoid voltage* across the yoke terminals.

You have now completed the programmed section.

ADDITIONAL PRACTICE CET TEST QUESTIONS

Some of the questions in this group were used in the past; or, they are contemplated for future use. Answers are given at the end of the chapter.

01. The sound carrier is above the video carrier when the television signal is transmitted. After the signal passes through the mixer the

(1) video carrier is still below the sound carrier.
(2) video carrier is above the sound carrier.

02. To replace a deflection yoke, a first consideration is the

(1) deflection angle.
(2) weight.
(3) size of wire used.
(4) type of core material.

03. Which of the following will decrease the brightness of the trace on a CRT?

(1) The cathode is made more positive.
(2) The cathode is made more negative.

04. An ion trap is not needed with

(1) aluminized picture tubes.
(2) a tube that has electrostatic deflection.
(3) a tube that has electromagnetic deflection.
(4) a tube that has electromagnetic focusing.

05. The white rectangle obtained when the electron beam moves *left and right,* and *up and down* is called the

(1) pincushion.
(2) getter.
(3) white square.
(4) raster.

06. In order to get a sawtooth current flowing through a magnetic deflection coil, the voltage across the coil will be a

(1) parabolic waveform.
(2) triangular waveform.
(3) sawtooth waveform.
(4) trapezoidal waveform.

07. A horizontal white line across the screen can be caused by

(1) a defective vertical oscillator.
(2) a shorted sync separator.
(3) a loss of high voltage.
(4) an open integrator.
(5) none of these.

08. The burst amplifier is keyed into conduction during

(1) monochrome signal reception only.
(2) the time the back porch of the horizontal blanking pedestal is present.
(3) the time when the VITS is present.
(4) the time when equalizing pulses are present.

09. The keying pulse goes to the keyed agc circuit from the

(1) vertical output stage.
(2) sync separator.
(3) flyback transformer.
(4) detector.

10. The boost voltage measures below normal in a certain tube-type receiver, but it rises to normal when the yoke is disconnected. Which is correct?

(1) This is an indication that the oscillator frequency is too low.

(2) This is an indication that there is a heater-to-cathode short in the horizontal output tube.
(3) This is an indication that the low-voltage supply has a defective filter.
(4) This is an indication that the yoke is defective.
(5) The luminance amplifier is defective.

11. If the horizontal retrace is not blanked the burst may show on the screen as

(1) a color strip on the right side of the screen.
(2) colored speckles of snow.

12. A blocking oscillator is most nearly like

(1) a Clapp oscillator.
(2) a Colpitts oscillator.
(3) an Armstrong oscillator.
(4) a multivibrator.
(5) a bridge oscillator.

13. Which of the following is least likely to be used as a sweep oscillator in a television receiver?

(1) A neon sawtooth oscillator
(2) A transistor blocking oscillator
(3) A transistor multivibrator

14. Which is correct?

(1) There are two frames per field.
(2) There are two fields per frame.

15. A keystone raster may be caused by a

(1) defective yoke.
(2) shorted burst amplifier input circuit.
(3) 120-hertz ripple in the horizontal sweep.
(4) 60-hertz ripple in the horizontal sweep.
(5) loss of the boost voltage.

16. Saturation is determined by the

(1) amplitude of the luminance signal.
(2) phase of the I and Q signals.
(3) amplitude of the color signal.
(4) frequency of the burst signal.
(5) the time for one cycle.

17. The automatic degaussing circuit of a color receiver employs a VDR and

(1) a varactor.

(2) a thermistor.
(3) an ABL.
(4) an LDR.
(5) an ALU.

18. Any color can be obtained with the proper amounts of hue, saturation, and

(1) phase.
(2) brightness.

19. Which of the following statements is true regarding the I and Q signals?

(1) The I and Q signals have the same frequency, and they are 180° out of phase.
(2) The I and Q signals have the same frequency, and they are 90° out of phase.
(3) The frequency of the I signal is higher than the frequency of the Q signal.
(4) The frequency of the I signal is lower than the frequency of the Q signal.
(5) (None of these statements is true.)

20. Two signal are in quadrature if they are

(1) equal in frequency.
(2) 90° out of phase.
(3) 180° out of phase.
(4) in phase.
(5) (none of these.)

21. The composite color input signal to the chroma bandpass amplifier is most likely to come from:

(1) the tuner.
(2) the mixer.
(3) the receiver synchronizing section.
(4) a receiver i-f amplifier.
(5) a video amplifier.

22. This keying pulse to the chroma bandpass amplifier

(1) increases the average dc level of the signal.
(2) removes the sound i-f signal.
(3) removes the color burst.
(4) removes the luminance signal.

23. The chrominance signal is separated from the composite color signal and fed to the demodulators in the

(1) bandpass amplifier.
(2) color killer.
(3) luminance amplifier.
(4) burst amplifier.
(5) sync separator.

24. The composite color signal does not include the

(1) chrominance and luminance signals.
(2) aft correction signal.
(3) synchronizing and blanking signals.
(4) burst signal.

25. A possible cause of color but no picture is

(1) a defective r-f amplifier.
(2) color-killer circuit not functioning properly.
(3) agc control not properly adjusted.
(4) open delay line.
(5) power supply fuse.

26. If the beam landing controls are not properly adjusted, it will affect

(1) only the color picture.
(2) both the monochrome and the color picture.
(3) only the monochrome picture.

27. The circuit that automatically disables the chrominance channel when no color signal is being received is the

(1) luminance circuit.
(2) acc circuit.
(3) abl circuit
(4) color-killer circuit.
(5) afc circuit.

28. The sound takeoff point in an intercarrier receiver is most likely to be at

(1) the rf amplifier.
(2) the first video amplifier.
(3) the i-f amplifier stage.
(4) the mixer stage.
(5) the sync separator.

29. In the i-f stages of a television receiver, traps are used for

(1) increasing the gain of the i-f amplifier.
(2) blocking sound signals.
(3) adjusting the shape of the i-f response curve.

(4) blocking rf signals.
(5) separating sync signals.

30. Which of the following types of detectors would NOT be used in the aft circuit of a fully transistorized TV receiver?

(1) Discriminator
(2) Quadrature
(3) Gated-beam detector
(4) Ratio detector
(5) Slope detector

31. In a line-operated television receiver, a zener diode is used in the B+ power supply

(1) for filtering.
(2) for voltage regulation.
(3) for rectifying.
(4) for voltage dividing.
(5) frequency control.

32. In a television receiver, the local-oscillator frequency can be varied by changing the

(1) abl adjustment.
(2) acc adjustment.
(3) fine tuning control.
(4) agc adjustment.
(5) video peaking control.

33. The maximum possible gain of the video i-f section is set by the

(1) trap adjustments.
(2) abl adjustment.
(3) acc adjustment.
(4) agc adjustment.
(5) afc adjustment.

34. The phase of the color subcarrier signal is varied by turning the

(1) agc control.
(2) optimizer control.
(3) tint control.
(4) color control.
(5) contrast control.

35. The aft correction voltage alters the

(1) chroma-amplifier bandpass.

(2) maximum CRT beam current.
(3) luminance delay time.
(4) local-oscillator frequency.
(5) color burst frequency.

36. Which of the following describes the hue of a color?

(1) Wavelength of the color
(2) Amount of white mixed with the color
(3) Brightness of the color

37. Which of the following is used for the NTSC reference white?

(1) Illuminant B
(2) Illuminant C
(3) Illuminant A

38. The subcarrier frequency for color is (approximately)

(1) 6 MHz.
(2) 4.25 MHz.
(3) 4.5 MHz.
(4) 3.58 MHz.
(5) 1.25 MHz.

39. The frame frequency is approximately

(1) 30 frames per second.
(2) 60 frames per second.
(3) 120 frames per second.
(4) 525 frames per second.
(5) 15,750 frames per second.

40. The color burst is used to

(1) present an amplitude-modulated color signal to the receiver for demodulation.
(2) supply a rainbow of colors for use in testing the receiver.
(3) supply bursts of color to the picture tube.
(4) synchronize the color oscillator in the receiver with the color subcarrier generator at the transmitter.

41. The scanning line frequency is approximately

(1) 60 lines per second.
(2) 30 lines per second.
(3) 15,750 lines per second.
(4) 525 lines per second.

42. The field frequency is approximately

(1) 15,750 fields per second.

(2) 60 fields per second.
(3) 525 fields per second.
(4) 30 fields per second.

43. An advantage of transmitting I and Q signals instead of transmitting R-Y and B-Y signals is

(1) more realistic blue sky.
(2) lower cost.
(3) more realistic green grass.
(4) more realistic flesh tones.
(5) greater distance for transmission.

44. Which of the following signals is transmitted with two complete sidebands?

(1) The I signal.
(2) The Q signal.

45. Which of the following will produce a change in hue?

(1) A change in the amplitude of the chrominance signal.
(2) A change in the phase of the chrominance signal.
(3) A change in the burst frequency.

46. The sweep frequency of your oscilloscope is 7.875 kHz, and you are observing the signal at the output of the first video amplifier. You should see

(1) two frames including two vertical blanking pedestals.
(2) two lines including two horizontal blanking pedestals.
(3) two fields, but no blanking pedestals.

47. The color burst is a minimum of eight cycles located

(1) at the top of the horizontal sync pulse.
(2) on the back porch of the horizontal blanking pedestal.
(3) in the 25 kHz guard band at the end of the channel.
(4) on the front porch of the vertical blanking pedestal.
(5) between equalizing pulses.

48. Which of the following statements is not true?

(1) The vertical sync pulse has a higher amplitude than the horizontal sync pulse.
(2) The vertical blanking period is longer than the horizontal blanking period.
(3) The horizontal lines are still being scanned during vertical retrace.
(4) Only about 480 lines out of a 525 line total contain video modulation.

(5) The starter circuit is required in a receiver with a scan-derived power supply.

49. The type of transmission used with television video signals is

(1) pulse position modulation.
(2) single-sideband, suppressed-carrier transmission.
(3) amplitude modulation with one complete sideband and one vestigial sideband.
(4) amplitude modulation with two sidebands.

50. In the transmitted television signal

(1) the video carrier is at a higher frequency than the center frequency of the sound signal.
(2) the video carrier is at a lower frequency than the center frequency of the sound carrier.

51. The number of active scanned lines for a frame is

(1) 125 to 250.
(2) 250 to 260.
(3) 470 to 480.
(4) 510 to 515.
(5) 500 to 510.

52. The video carrier of a composite signal is

(1) 0.25 MHz below the upper end of the channel.
(2) 0.25 MHz above the lower end of the channel.
(3) 1.25 MHz below the upper end of the channel.
(4) 1.25 MHz above the lower end of the channel.

53. The maximum deviation of the sound signal for television is

(1) 10 kHz.
(2) 25 kHz (a total carrier swing of 50 kHz).
(3) 75 kHz (a total carrier swing of 150 kHz).
(4) 80.8 kHz.
(5) 100 kHz (a total carrier swing of 200 kHz).

54. The center frequency of the fm sound signal and the video carrier frequency are

(1) 4.5 MHz apart.
(2) 4.5 kHz apart.
(3) 45 MHz apart.
(4) 45 kHz apart.
(5) none of these choices is correct.

55. The ideal response curve of a television receiver causes the video signal at the carrier to be reduced in amplitude by 50 percent. This is required in order to

(1) prevent overemphasis of lower video frequencies due to the fact that they are transmitted as a vestigial sideband.
(2) reduce cross talk between the video carrier and the sound carrier of the channel being received.
(3) reduce cross talk between the video carrier and the sound carrier of the adjacent channel.

56. Select the correct statement.

(1) Audio is transmitted single sideband in the television signal.
(2) The field frequency is 1/500 the horizontal frequency.
(3) The horizontal sweep frequency is 500 times the vertical sweep frequency.
(4) The field frequency is one-half the frame frequency.
(5) The field frequency and the vertical sweep frequency are the same value.

57. The selectivity and sensitivity of a tuner is governed primarily by

(1) the rf amplifier stage.
(2) the method of selecting tuned circuits.
(3) the mixer stage.
(4) the method of coupling between the rf amplifier and mixer.
(5) the local oscillator stage.

58. Which of the following types of detectors would not be used in the aft circuit of a fully transistorized receiver?

(1) Discriminator.
(2) Ratio detector.
(3) Gated-beam detector.

59. A limiter circuit is associated with

(1) discriminators.
(2) gated-beam detectors.
(3) de-emphasis circuits.
(4) ratio detectors.

60. The signal for vertical-retrace blanking goes to

(1) the video i-f amplifier.
(2) the video detector.
(3) the video amplifier.

(4) the agc circuit.
(5) the aft circuit.

61. The top of the picture is stretched out. You should first

(1) replace the yoke.
(2) adjust the vertical linearity control.

62. A certain receiver will not hold sync horizontally or vertically. Other than this, the picture looks normal. A likely cause is

(1) a defective yoke.
(2) a defective sync separator.

ANSWERS TO PRACTICE TEST

Question Number	Answer Number
01	(2)
02	(1)
03	(1)
04	(1)
05	(4)
06	(4)
07	(1)
08	(2)
09	(3)
10	(4)
11	(1)
12	(3)
13	(1)
14	(2)
15	(1)
16	(3)
17	(2)
18	(2)
19	(2)
20	(2)
21	(5)
22	(3)
23	(1)
24	(2)
25	(4)
26	(2)
27	(4)
28	(2)
29	(3)
30	(3)
31	(2)
32	(3)
33	(4)
34	(3)
35	(4)
36	(1)
37	(2)
38	(4)
39	(1)

Question Number	Answer Number
40	(4)
41	(3)
42	(2)
43	(4)
44	(2)
45	(2)
46	(2)
47	(2)
48	(1)
49	(3)
50	(2)
51	(3)
52	(4)
53	(2)
54	(1)
55	(1)
56	(5)
57	(1)
58	(3)
59	(1)
60	(3)
61	(2)
62	(2)

Test Equipment and Troubleshooting

Section 15 of the test is called *Test Equipment.* It contains 5 questions. Section 14 is called *Troubleshooting Consumer Equipment,* and it contains 15 questions. This chapter reviews the types of questions you will find in these two sections.

Traditionally technicians do very well in this part of the CET test. In the first place, anyone taking this part of the test is presumed to have four years of equivalent practical experience. So, he is very familiar with the equipment and test procedures used for television servicing.

It is important for you to understand the distinction made in the CET test between *Tests and Measurements* and *Troubleshooting.* Tests and measurements include the setups, adjustments, and tests and measurements that you might make in preparation for troubleshooting.

For example, you might be asked how it is possible to obtain the response curve for the color bandpass amplifier. This is actually a test setup. Or, you may be asked a question on how to measure the height of a sync pulse with an oscilloscope. That would be a measurement.

In the measurement section you would not be asked what might be the possible cause of a symptom such as the yellow line down the side of a color picture. That would be a troubleshooting question.

You will get a good idea of the types of questions in each section in the practice test in Appendix A. To answer questions in troubleshooting, a technician must be able to recognize symptoms, and must be able to perform a logical troubleshooting procedure to locate a defective component or part.

In the actual test you may be shown the schematic of a complete system and asked the cause of incorrect voltage or waveform measurements. Actually, such questions are no different from questions about individual circuit problems.

In some cases you may be asked about making repairs. For example, you may be asked about the type of solder used for replacing devices, the use of heatsinks, or the use of substitute replacement parts. Expect questions on troubleshooting logic circuits with a logic probe, and about the use of oscilloscope and voltmeter probes.

BASIC METER MOVEMENTS

A galvanometer is a very sensitive current measuring device. It is not uncommon to have galvanometers with full scale deflection of 25 microamperes or 50 microamperes. The *deflection sensitivity* of a meter is the reciprocal of its full-scale deflection. This sensitivity is given in ohms per volt. Mathematically:

$$\text{ohms-per-volt rating} = \frac{1}{\text{full scale current}}$$

Note this very important point: *the ohms-per-volt rating of a meter has nothing to do with the number of ohms impedance the meter offers to the circuit when making a voltage measurement.*

Typical CET Test Question

A certain meter is rated at 100,000 ohms per volt. This means that

(1) the impedance of the meter is 100,000 ohms for each volt being measured.
(2) the full scale deflection of the meter movement is 10 microamperes.

Answer: Choice (2) is correct. The reciprocal of 100,000 ohms per volts is 0.00001 ampere. If you take the reciprocal of *ohms-per-volt* you get *volts-per-ohm*. According to Ohm's Law amperes is equal to volts per ohm. Therefore, the reciprocal of the ohms-per-volt reading is a *current* value.

Meter movements can be divided into two basic groups—those

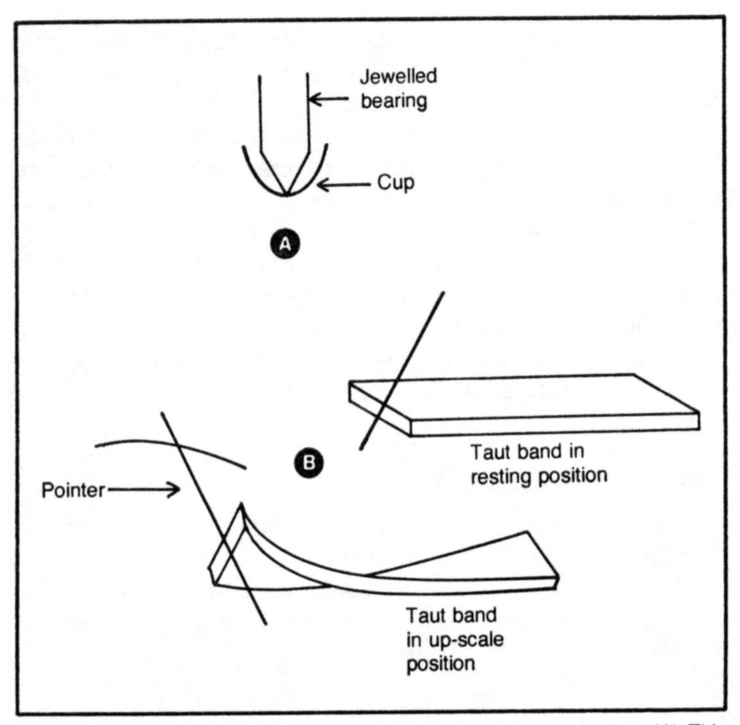

Fig. 8-1. Analog meter movements are made in two basic designs (A) This jeweled bearing offers very little friction. (B) The taut band provides the restoring force for the pointer.

with jeweled bearings and those that use a taut band. Figure 8-1 compares these two movements. Figure 8-1A shows the side view of a jeweled bearing. Note that the bearing fits inside of a wedge-shaped cup. As the meter moves upscale this bearing turns in the cup. Hopefully, it will turn with a minimum amount of friction. When the meter is new there is very little problem here; but, as the meter gets older certain basic disadvantages of this type of movement become apparent.

One disadvantage is that the bearing eventually begins to wear and no longer fits with the minimum contact area in the cup. When that happens, the meter begins to stick at various points. For this reason, technicians are advised to tap the meter gently when making a measurement with a meter that uses this type of movement.

It should be obvious if you stand this type of meter on end the jewel is going to ride against the bottom of the cup rather than in the center of the cup. Therefore minimum friction occurs when the meter is laid on its back rather than when it is standing up.

If the jeweled-type meter movement is subjected to a strong jolt (which may occur from dropping it) permanent damage can occur to this simple movement and the meter will no longer be useful.

The taut band meter movement employs the twist of a band as the meter is moved upscale. Figure 8-1B shows the taut band in its resting position, and the band as the meter is moved upscale. This band is used for the restoring force of the pointer. It moves it back to zero when the measurement is completed. In the jeweled type a small spring is used for the restoring force. The taut band movement is considered to be much more reliable and much more rugged. It is generally found in the more expensive meter movements.

Remember that a meter movement is converted to a voltmeter by connecting a *series multiplier* in series with the meter movement. To convert the meter movement to a current-reading instrument (such as a milliameter or ammeter) a *shunt resistor* is connected across the meter movement terminals. You will not be required to calculate the value of a multiplier or a shunt in the CET test, but you should understand their function.

To convert the meter movement to an ohmmeter it is necessary to supply a battery voltage and either a series multiplier or a shunt to protect the meter during the measurement. Since the battery supplies a current to the circuit being measured, you must be careful not to overload sensitive diodes and transistor junctions when measuring their resistance. This is most likely to occur when the meter is in the R × 1 position.

Digital meters are much more accurate and convenient to use than analog types. Furthermore, they do not have the problem of parallax that is present in the analog type. Parallax occurs when you view the meter pointer from an angle rather than from directly above.

To reduce the problem of parallax, some analog meters have a *mirrored scale*. Its purpose is to assure that the measurement is taken with the viewer directly above the meter pointer. The technique is to view the pointer in such a way that you do not see its reflection in the mirror. That can only occur if you're directly above the pointer. From that position the measurement is made.

OSCILLOSCOPES

As a general rule triggered-sweep scopes are preferred over recurrent sweep scopes for television servicing. One advantage of the triggered-sweep scope is that the sweep can be calibrated very accurately in microseconds (or milliseconds) per centimeter (or inch). This makes it very easy to find the time (T) for one cycle of an input waveform. By using the equation

$$f = \frac{1}{T}$$

it is possible to calculate the frequency of the signal being displayed.

In order to calculate the frequency of the display with the recurrent sweep scope it is necessary to use some type of marker. One way is to apply very narrow marker pulses to "Z" axis. *Remember that the "Z" axis of the oscilloscope controls the brightness of the display.*

Another advantage of the triggered-sweep scope is that it will stay locked onto the incoming signal even if the frequency of that signal varies slightly. With the recurrent sweep scope it is necessary to readjust the sweep frequency for even a small change in the frequency of the signal being displayed.

Be sure that you can calculate the frequency ratios of two sine waves in a Lissajous pattern. The simplest way to do this is illustrated in Fig. 8-2. The number of times that the display touches the horizontal line is directly related to the frequency of the vertical signal, or, the frequency on the vertical terminals. The number of times that the display touches the vertical line is related to the frequency of the signal being applied to the horizontal input terminals. Remembering this simple fact you can write the equation

$$\frac{V}{H} = \frac{\text{no. of times display touches the horizontal line}}{\text{no. of times display touches the vertical line}}$$

Typical CET Test Question

What is the ratio of vertical to horizontal frequency for the Lissajous pattern being displayed in Fig. 8-2?

Answer: the pattern touches the horizontal line three places and touches the vertical line two places. Putting these into the equation just given gives you

$$\frac{V}{H} = \frac{3}{2}$$

The result shows that the vertical frequency is 3/2 times the

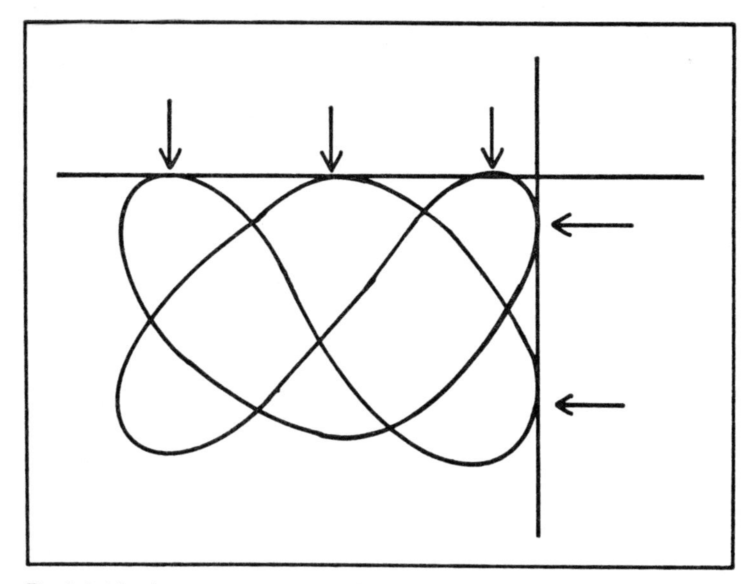

Fig. 8-2. Lissajous patterns are used for comparing frequencies.

horizontal frequency. So, if the horizontal frequency is 300 Hertz, the vertical frequency will be:

$$\text{vertical frequency} = \frac{3}{2} \times 300 = 450 \text{ Hertz}$$

The Lissajous pattern can also be used to tell the phase difference between two sine-wave signals. The technique is based on the measurements shown in Fig. 8-3. Using these measurements the equation for the phase angle is given as follows:

$$\text{phase angle } (\theta) = \text{arc sine } \frac{A}{B} = \text{Sin}^{-1} \frac{A}{B}$$

The term arc sine simply means the angle that has a sine equal to that ratio. For example the arc sine of 0.5 is 30°. The arc sine on calculators is usually called

$$\text{SIN}^{-1}$$

Remember that a Lissajous pattern comparing two sine waves with frequencies that are equal will produce a perfect circle if the two sine waves are 90° out of phase. If they are 0° out of phase there will be a straight line rising from left to right. If they are 180° out of phase there will be a straight decreasing from left to right.

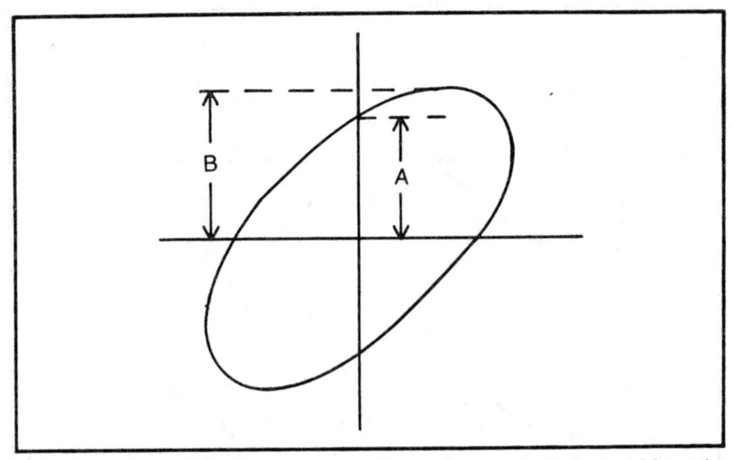

Fig. 8-3. The phase angle between two sine waves can be calculated from the measurements shown here.

Typical CET Test Question

A Lissajous pattern is obtained by comparing the input sine-wave signal to output sine-wave signal for a class A amplifier. This

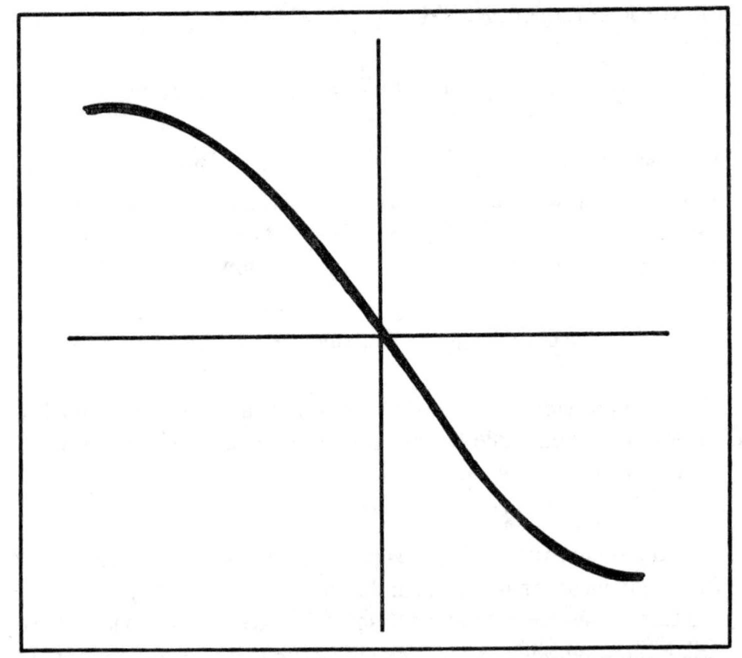

Fig. 8-4. This Lissajous pattern indicates that there is non-linear distortion in the amplifier.

should produce a straight line that rises as it moves to the left. However, for one particular amplifier the line is not straight, but rather, appears as shown in Fig. 8-4. What is the indication?

(1) The output frequency does not equal the input frequency.
(2) The signals are not exactly 180° out of phase.
(3) The amplifier has some form of linear distortion.
(4) None of these choices is correct.

Answer: Choice (3) is correct. If there is a line, but it is not straight, the indication is that some form of nonlinear distortion is occurring in the amplifier.

Remember that you can use an oscilloscope to measure current. The technique is to place a one ohm or ten ohm resistor in series with the current being measured. The scope is then used to measure the voltage across that resistor.

Triggered-sweep scopes are generally designed with an internal calibration for voltage measurements. To use a recurrent sweep scope for measuring the voltage it is necessary to use some form of the voltage calibrator to assist in making the measurement.

It is sometimes necessary to use special probes for making measurements with an oscilloscope or voltmeter. Most inexpensive oscilloscopes cannot display the intermediate frequency of a basic AM receiver let alone the intermediate frequency of a television receiver. To get around this problem a detector probe is normally used. It shows the envelope of the intermediate frequency signal, but not the intermediate frequency signals that make up that envelope. Presumably, if the envelope is not correct the i-f stage is not operating properly.

High-voltage probes are used with voltmeters. They are generally marked with the amount of multiplication necessary to convert the voltmeter reading to the actual value being made. For example if a times 10 probe is used to make a measurement, and the voltmeter indicates a hundred volts, the actual volt being measured is 100 × ten or 1000 volts.

Be sure you are familiar with the square wave test of an amplifier. In this test a square wave is applied to the input of the amplifier and the output is observed on an oscilloscope. This test should never be used with an amplifier that has an inductive load. There are two reasons for this. One is that the inductive kickback could destroy the transistor unless the designer has made some special provision against this possibility. The second reason is that the reactive load will change the shape of the square wave and may give an erroneous indication of poor performance.

The most important output square wave patterns to know are illustrated in Fig. 8-5. Observe that an overshoot on the square waveform is an indication of *excessive high frequency response*, but it also may be an indication of an inductive load. This characteristic is utilized in the ringing test of test equipment designed to evaluate yokes, transformers and other inductive components. What happens with this type of test equipment is that the inductor is electrically shock excited with a pulse. Assuming that the inductor is not short circuited it should go into self oscillation with the distributed capacitance. The result is a number of cycles of ringing. However, if the inductor is defective only one or two (or maybe three) cycles of ringing will be observed. The ringing test is a quick way to evaluate

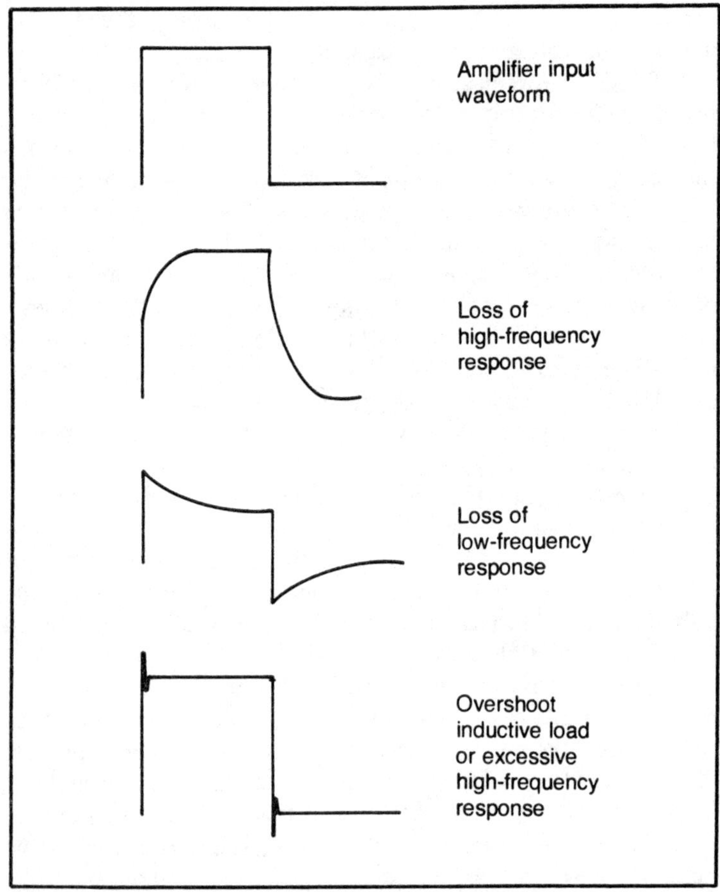

Fig. 8-5. You are likely to be asked to identify the results of a square-wave test. The most important ones are shown here.

the ability of the inductor to perform. If you want the inductance *value*, it should be measured on a bridge, or (indirectly) on a Q-meter.

EVALUATING PARAMETERS

As a technician you should be very familiar with the methods used to evaluate tubes, bipolar transistors, and FETs. The various parameters are defined in Table 8-1. Pay particular attention to similarities between the FETs and the tubes. Note also the difference between the ac and dc parameters used for measuring transistor performance. Also, note that alpha (h_{FB}) is a parameter obtained with the transistor in the *common base* configuration while beta (h_{FE}) is a parameter that is obtained with the transistor in the *common emitter* configuration.

A quick test for transistor operation is to use an ohmmeter with the procedure described in Fig. 8-6. You should be careful not to use the ohmmeter on the R×1 scale of a volt-ohm-milliammeter (VOM) because that can readily burn out a diode or transistor junction.

TESTING AMPLIFIERS

A quick test that is sometimes used for determining if a transistor amplifier is operating is illustrated in Fig. 8-7. In this test the emitter-to-base junction is shorted while the collector voltage is observed. Theoretically, at least, shorting the emitter to the base of the transistor does not hurt it. It simply removes the forward bias, and therefore, prevents the transistor from conducting.

If the transistor shuts off with emitter-base junction shorted it means that there has been control over the collector current by the transistor base. Obviously, if the collector current stops flowing the collector voltage in the illustrated circuit should go to the +dc value.

This test is somewhat controversial because in the hands of an inexperienced technician some erroneous data can be obtained. For example, on transistors that are direct coupled the bias of one transistor is obtained directly from the collector of the previous transistor. Shutting off the previous transistor causes the collector voltage to go to a highly positive value. That, in turn, will destroy the second transistor in the direct-coupled configuration.

Another problem with the short-circuit quick test of a transistor amplifier is that some transistors are operated in class B configuration. Remember that in a class B configuration the bias between the emitter and the base is 0 volts (or nearly so). Shorting the

Table 8-1. Bipolar Transistor, FET, and Vacuum-Tube Parameters.

NOTE: As a general rule, the dc characteristics are noted with capital letters and the ac (or "dynamic") characteristics are noted with lower-case letters. The symbol Δ (delta) means "a small change in value." Thus, ΔE_g means a small change in grid voltage.

	Name	Symbol	Description	Equation
TUBES	Amplification Factor	μ	Measure of the ability of the tube to amplify	$\mu = -\frac{\Delta E_p}{\Delta E_g}\Big]\ I_p =$ Constant
	Plate Resistance	r_p	Plate-to-cathode resistance of the tube	$r_p = \frac{\Delta E_p}{\Delta I_p}\Big]\ E_g =$ Constant
	Transconductance (Also known as mutual conductance)	g_m	Change in plate current produced by a change in grid voltage	$g_m = \frac{\Delta I_p}{\Delta E_p}\Big]\ E_p =$ Constant
	Relationship between μ, g_m and r_p			$\mu = g_m r_p$
BIPOLAR TRANSISTORS	Dc Alpha	α_{DC}	Common-base forward current transfer ratio	$\alpha_{DC} = \frac{I_C}{I_E}$
		h_{FB}	Same as dc alpha	$h_{FB} = \alpha$

	Dc Beta	β_{DC}	Common-emitter forward-current transfer ratio	$\beta_{DC} = \frac{I_C}{I_B}$
		h_{FE}	Same as dc beta	
	Gain-bandwidth Product	f_T	Frequency at which $\beta = 1$	$h_{FE} = \beta$
	Ac Alpha	α_{ac}	Common-base forward-current transfer ratio using signals or current changes (instead of dc currents as in α_{dc})	$\alpha = h_{fb} = \frac{\Delta I_C}{\Delta I_E} = \frac{i_c}{i_3}$
	Ac Beta	β_{AC}	Common-base forward-current transfer ratio using signals or current changes (instead of dc currents as in β_{DC})	$\beta = h_{fe} = \frac{\Delta I_C}{\Delta I_B} = \frac{i_c}{i_b}$
	Relationship between α and β			$\alpha = \frac{\beta}{1 + \beta}$ $\beta = \frac{\alpha}{1 - \alpha}$
FETs	Transconductance	g_m	Similar to the transconductance of vacuum tubes	$g_m = \frac{\Delta I_D}{\Delta V_{GS}}$ $V_{DS} = Constant$

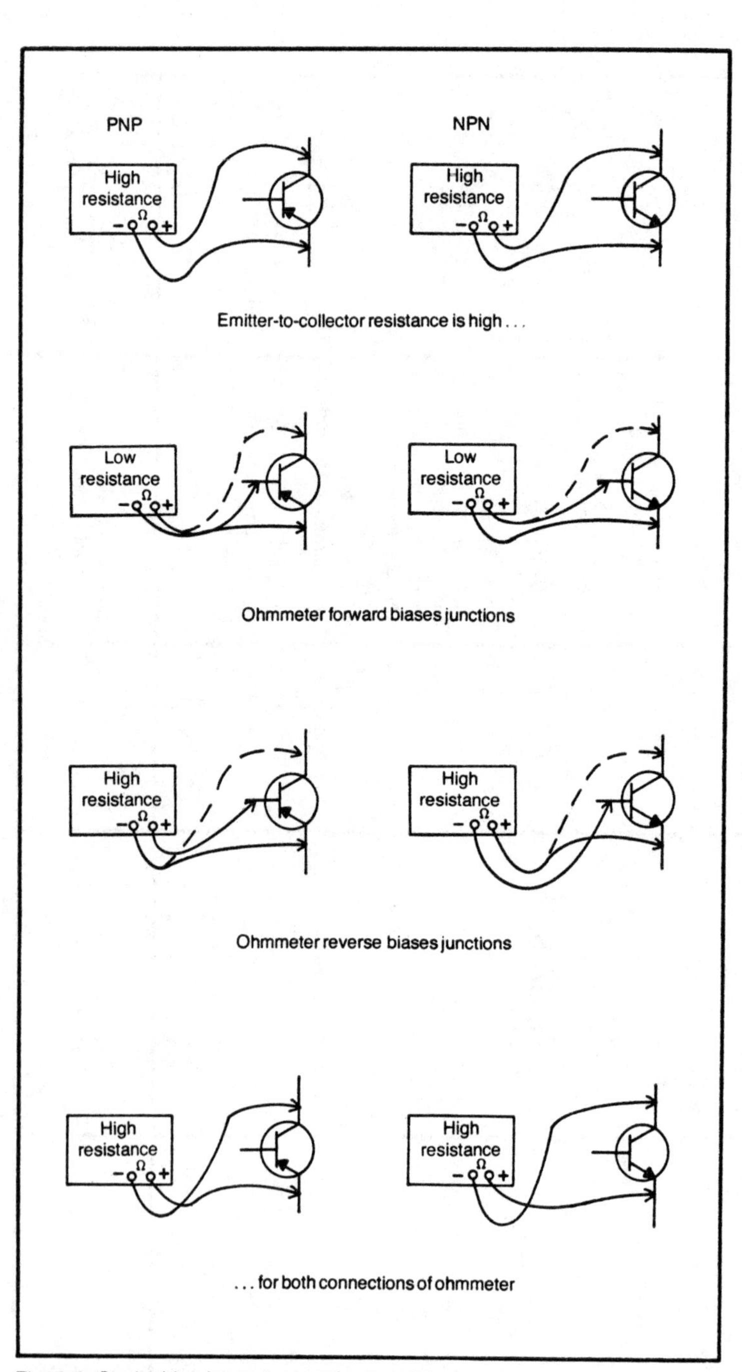

Fig. 8-6. Study this ohmmeter test for transistors.

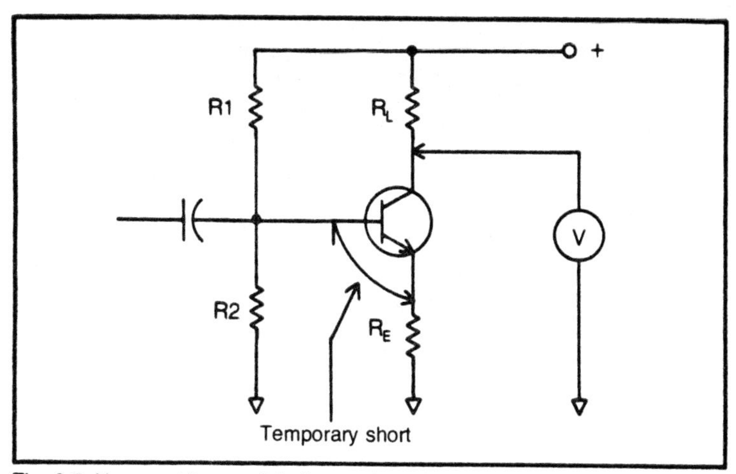

Fig. 8-7. You must be careful when making this short-circuit test!

emitter to the base in that configuration will not produce any observable change in the collector voltage. As a matter of fact the collector voltage will already be at its maximum positive value in the absence of an input signal.

Transistors in push pull and totem pole configurations are normally operated in a class AB configuration. The reason for this is that a small amount of forward bias is necessary to eliminate the *crossover distortion* that occurs when one transistor stops operating and the other one begins. The forward bias obtained with class AB operation prevents one transistor from reaching cutoff before the other transistor begins to conduct.

This type of distortion is easily observed by applying a pure sine wave to an improperly-biased push pull or totem pole configuration and observing the output across the load. Figure 8-8 shows the output signal with crossover distortion.

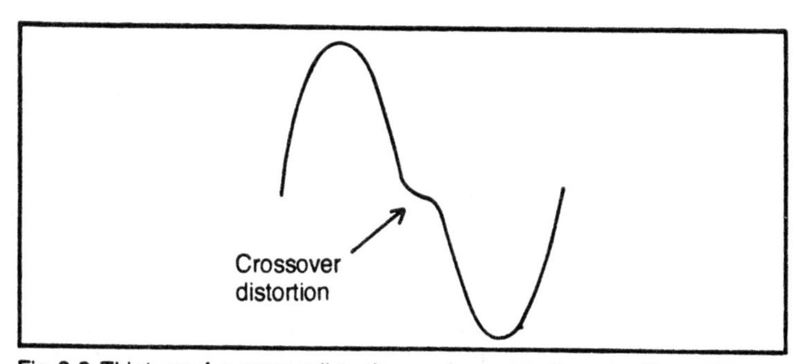

Fig. 8-8. This type of crossover distortion can be avoided by using Class AB bias.

In order to properly evaluate a collector voltage measurement it is necessary for you to understand the transistor configuration. Specifically the circuit must be viewed in relation to the power supply voltage.

Figure 8-9 shows two ways to connect an npn transistor across a dc power supply. Observe the differences in voltmeter readings when the short circuit test is applied. This is another criticism of the short circuit test used for transistor amplifiers. However, despite the various disadvantages that have been discussed so far, ***the short***

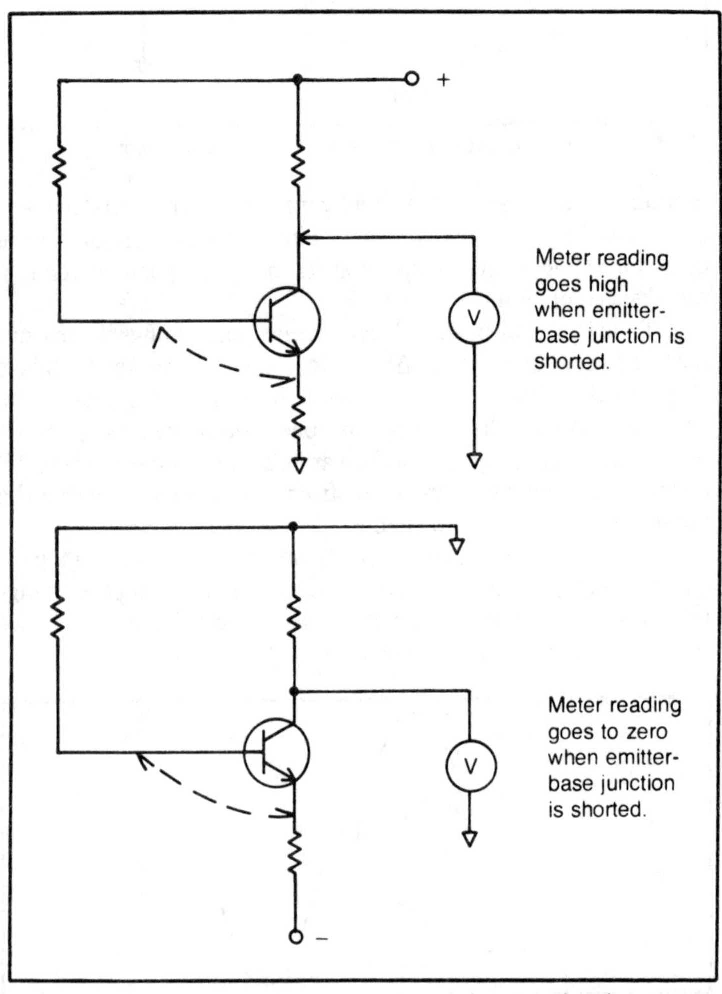

Fig. 8-9. The power supply voltage can be delivered to a transistor in two ways. Note how the short-circuit test is affected.

circuit test is considered to be a valuable troubleshooting technique and it is used by many technicians in the field. As a matter of fact, some pieces of equipment in the field are marked on the printed circuit board to show the technician where the short circuit test can be safely applied. This relieves the technician of the responsibility of making sure that the proper configuration is being tested.

PROGRAMMED REVIEW

Start with block number 1. Pick the answer that you feel is correct. If you select choice number 1, go to Block 13. If you select choice number 2, go to Block 15. If you select choice number 3, go to Block 4. Proceed as directed. There is only one correct answer for each question.

BLOCK 1

The rise time of a pulse is defined as the time it takes the voltage to go from

(1) its minimum value to its maximum value. Go to Block 13.
(2) 10% of maximum to 90% of maximum. Go to Block 15.
(3) neither choice is correct. Go to Block 4.

BLOCK 2

Your answer to the question in Block 15 is not correct. Go back and read the question again and select another answer.

BLOCK 3

The correct answer to the question in Block 38 is choice (2). The method shown in X is called signal injection.

Here is your next question:

Power transistors in a push-pull amplifier are operated class AB in order to prevent

(1) speaker overload. Go to Block 29.
(2) crossover distortion. Go to Block 7.

BLOCK 4

Your answer to the question in Block 1 is not correct. Go back and read the question again and select another answer.

BLOCK 5

Your answer to the question in Block 20 is not correct. Go back

and read the question again and select another answer.

BLOCK 6

Your answer to the question in Block 37 is not correct. Go back and read the question again and select another answer.

BLOCK 7

The correct answer to the question in Block 3 is choice (2). Note that the amplifiers are not operated class B.

Here is your next question:

Which of the following is a better test for the deflection yoke of a picture tube?

(1) Ringing test. Go to Block 10.
(2) Ohmmeter test. Go to Block 34.

BLOCK 8

The correct answer to the question in Block 27 is choice (2). The amplifier is a common emitter configuration. The configuration is not determined by the dc voltages.

Here is your next question:

In a color television receiver it might be necessary to align the

(1) luminance amplifier. Go to Block 12.
(2) bandpass amplifier. Go to Block 19.

BLOCK 9

Your answer to the question in Block 16 is not correct. Go back and read the question again and select another answer.

BLOCK 10

The correct answer to the question in Block 7 is choice (1). If the coil has a few shorted turns the ohmmeter test is the next thing to useless.

Here is your next question:

To convert a recurrent sweep oscilloscope to dual trace, you would use

(1) an electronic switch. Go to Block 27.
(2) an electronic doubler. Go to Block 23.

BLOCK 11

The correct answer to the question in Block 19 is choice (1).

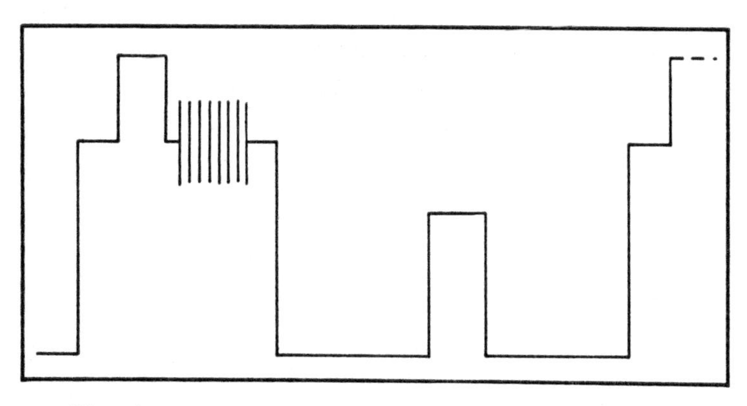

Here is your next question:

Consider the waveform shown in this block. If the video signal is the same as shown for every line, the screen will show

_______________. Go to Block 39.

BLOCK 12

Your answer to the question in Block 8 is not correct. Go back and read the question again and select another answer.

BLOCK 13

Your answer to the question in Block 1 is not correct. Go back and read the question again and select another answer.

BLOCK 14

The correct answer to the question in Block 24 is choice (2). The RMS value is only 0.707 times the peak value for a pure sine wave. Likewise, the average value is only 0.636 times the peak value for a pure sine wave.

Here is your next question:

To measure the emitter-to base voltage of a PNP transistor, connect the positive lead of the voltmeter to the

(1) transistor base. Go to Block 32.
(2) transistor emitter. Go to Block 28.

BLOCK 15

The correct answer to the question in Block 1 is choice (2). That is the definition of rise time.

Here is your next question:

To check the calibration of a milliammeter,

(1) connect it in series with one that is known to be good and measure a current in the range of the instrument. Go to Block 24.
(2) Connect it in parallel with one that is known to be good and measure a current in the range of the instrument. Go to Block 2.

BLOCK 16

The correct answer to the question in Block 21 is choice (1). This is *not* a preferred way to check fuses. However, if the fuse can be easily removed it is a good way to measure the circuit current.

Here is your next question:

Bandwidth is the range of frequencies between points where

(1) the power drops to 70.7% of maximum. Go to Block 9.
(2) the voltage drops to 70.7% of maximum. Go to Block 37.

BLOCK 17

Your answer to the question in Block 28 is not correct. Go back and read the question again and select another answer.

BLOCK 18

Your answer to the question in Block 38 is not correct. Go back and read the question again and select another answer.

BLOCK 19

The correct answer to the question in Block 8 is choice (2). There are no alignment adjustments in a luminance amplifier. A possible exception would be a trap, but adjusting that trap is not a typical alignment procedure.

Here is your next question:

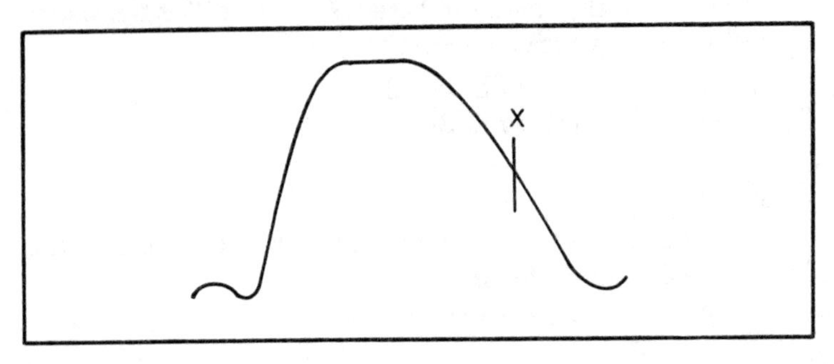

An ideal i-f response curve for a television receiver is shown here. The frequency marked 'X' locates the

(1) video carrier. Go to Block 11.
(2) sound carrier. Go to Block 35.

BLOCK 20

The correct answer to the question in Block 28 is choice (2). The 'Z' axis can be used for putting markers on the trace.

Here is your next question:

Resistor R3 is open in the circuit shown in this block. The voltmeter should read

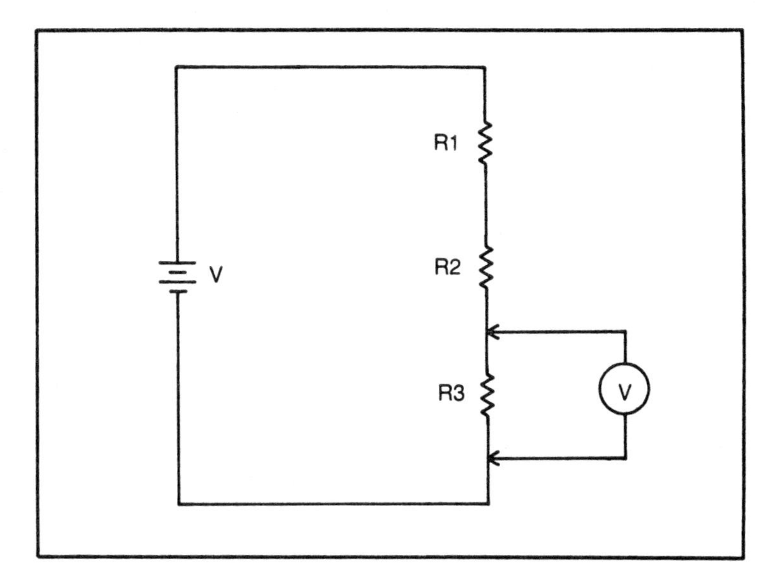

(1) zero volts. Go to Block 5.
(2) 12 volts. Go to Block 21.

BLOCK 21

The correct answer to the question in Block 20 is choice (2). All of the voltage should appear across the open circuit.

Here is your next question:

A milliammeter is connected across a fuse in a dc circuit. It displays a reading of 175 milliamperes. This indicates

(1) the fuse is open. Go to Block 16.
(2) the fuse is OK. Go to Block 30.

BLOCK 22

Your answer to the question in Block 24 is not correct. Go back and read the question again and select another answer.

BLOCK 23

Your answer to the question in Block 10 is not correct. Go back and read the question again and select another answer.

BLOCK 24

The correct answer to the question in Block 15 is choice (1). The meters are connected in series so the same amount of current flows through each meter. Ideally, the range of current values could be adjusted while the meter is being checked.

Here is your next question:

An oscilloscope shows that the peak-to-peak voltage of a sawtooth waveform is 10 volts. The RMS value of the voltage

(1) is 7.07 volts. Go to Block 22.
(2) cannot be determined from the information given. Go to Block 14.

BLOCK 25

The correct answer to the question in Block 37 is choice (1).

Here is your next question:

Which of the following is correct regarding the gate-to-source voltage of an N-channel enhancement MOSFET? (Assume it is working as a Class A amplifier.)

(1) The gate should be positive with respect to the source. Go to Block 38.
(2) The gate should be negative with respect to the source. Go to Block 26.

BLOCK 26

Your answer to the question in Block 25 is not correct. Go back and read the question again and select another answer.

BLOCK 27

The correct answer to the question in Block 10 is choice (1).

Here is your next question:

In the circuit shown in this block, the output signal should be

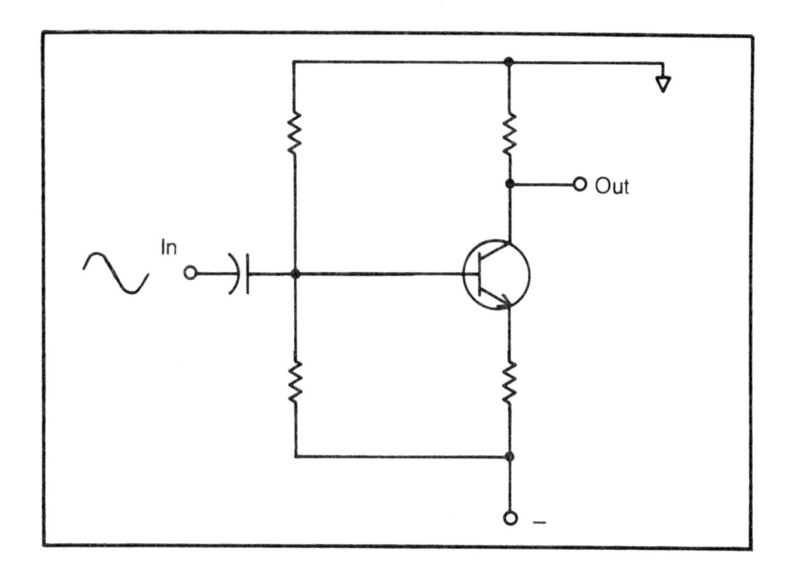

(1) in phase with the input signal. Go to Block 31.
(2) 180° out of phase with the input signal. Go to Block 8.

BLOCK 28

The correct answer to the question in Block 14 is choice (2). The base will be negative with respect to the emitter.

Here is your next question:

The brightness of an oscilloscope trace can be controlled with a voltage delivered to the

(1) B axis input. Go to Block 17.
(2) Z axis input. Go to Block 20.

BLOCK 29

Your answer to the question in Block 3 is not correct. Go back and read the question again and select another answer.

BLOCK 30

Your answer to the question in Block 21 is not correct. Go back and read the question again and select another answer.

BLOCK 31

Your answer to the question in Block 27 is not correct. Go back and read the question again and select another answer.

BLOCK 32

Your answer to the question in Block 14 is not correct. Go back and read the question again and select another answer.

BLOCK 33

Your answer to the question in Block 38 is not correct. Go back and read the question again and select another answer.

BLOCK 34

Your answer to the question in Block 7 is not correct. Go back and read the question again and select another answer.

BLOCK 35

Your answer to the question in Block 19 is not correct. Go back and read the question again and select another answer.

BLOCK 36

Your answer to the question in Block 38 is not correct. Go back and read the question again and select another answer.

BLOCK 37

The correct answer to the question in Block 16 is choice (2). Bandwidth is also measured between the half power points. Here is your next question:

A good technique for troubleshooting a closed-loop system is to

(1) open the loop and substitute a voltage or frequency as required. Go to Block 25.
(2) short circuit the output of the loop and see if the current in the feedback decreases. Go to Block 6.

BLOCK 38

The correct answer to the question in Block 25 is choice (1). Do not confuse enhancement MOSFET's with the depletion type.

Here is your next question:

Assume that measurements are made by moving from a to b to c. Which of the methods in this block illustrates the signal tracing procedure for troubleshooting?

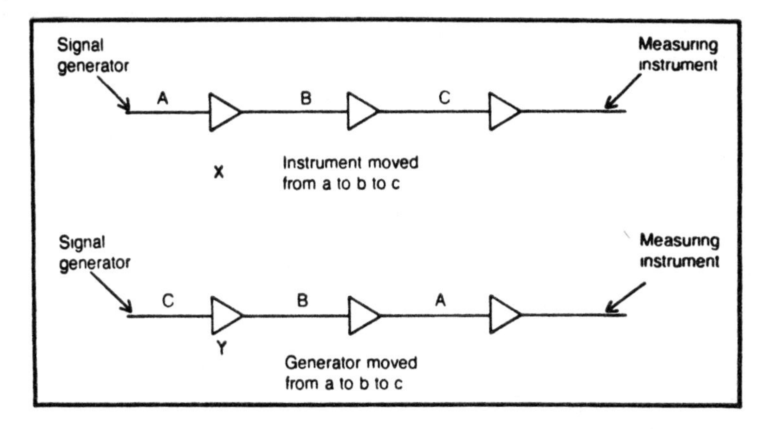

(1) Neither. Go to Block 33.
(2) Both. Go to Block 18.
(3) The one marked 'X'. Go to Block 3.
(4) The one marked 'Y'. Go to Block 36.

BLOCK 39

The correct answer to the question in Block 11 is that there will be a dark vertical line in the center of the screen.

You have now completed the programmed section.

ADDITIONAL PRACTICE CET TEST QUESTIONS

Some of the questions in this group were used in the past; or, they are contemplated for future use. Answers are given at the end of the chapter.

01. Which of the following statements is correct?

(1) The forward bias current for a transistor that is operating Class B will be greater than the forward bias on a transistor that is operating Class A.
(2) The forward bias current for a transistor that is operating Class A will be greater then the forward bias on a transistor that is operating Class B.

02. Which of the following is true?

(1) Shorting the base to the emitter is almost sure to destroy the transistor if it is in a Class-A amplifier circuit.
(2) Shorting the base to the collector is almost sure to destroy the transistor.

03. To check a transistor with a ohmmeter, set the ohmmeter on the

(1) R × 1 meg position.
(2) R × 100 position.
(3) R × 1 position.

04. Which of the following is not correct for ac alpha?

(1) $h_{fe} = \frac{\Delta I_c}{\Delta I_b}$

(2) $h_{FE} = \frac{\Delta I_c}{\Delta I_b}$

05. In the normal operation of a transistor, which junction is forward biased?

(1) The collector-base junction
(2) The emitter-base junction

06. The reverse current flow in the collector-base junction, with the emitter open is called

(1) I_{CER}.
(2) I_{CEO}.
(3) I_{ECS}.
(4) I_{CBO}.

07. To determine the reverse resistance of an NPN collector-base junction,

(1) place the positive lead of the ohmmeter on the collector terminal and the negative lead on the base terminal.
(2) place the negative lead of the ohmmeter on the collector terminal and the positive lead on the base terminal.

08. The bandpass amplifier must have a flat response for a range of

(1) 0.5 MHz on both sides of the color subcarrier.
(2) 3.58 MHz on both sides of the color subcarrier.
(3) 0.5 MHz above the color subcarrier and 4.5 MHz below the color subcarrier.
(4) 4.5 MHz on both sides of the color subcarrier.

09. For correct flesh tones, the tint control should be

(1) in its midrange position.
(2) at its maximum counterclockwise position.
(3) in its maximum clockwise position.

10. The output voltage of the video detector in a television receiver should be measured with

(1) a VTVM to determine the rms value of the signal.
(2) an oscilloscope to determine the peak-to-peak voltage of the video signal.

11. Which of the following represents a typical frequency range for a color bandpass amplifier?

(1) 0 - 1.2 MHz
(2) 1.0 - 4.5 MHz
(3) 3.0 - 4.1 MHz
(4) 4.1 - 4.5 MHz

12. To make a quick check of the high-frequency response of a video-amplifier stage, use

(1) a sweep generator with markers.
(2) an rf generator and a VOM.
(3) a color generator.
(4) a square-wave generator.

13. One difference between a triggered-sweep scope and a recurrent sweep oscilloscope is that

(1) the triggered-sweep scope does not have a trace.
(2) the triggered-sweep scope cannot be used for voltage measurements.

14. Which of the following explains the use of markers during receiver alignment.

(1) They blank the scope during retrace.
(2) They identify certain amplitudes on the response curves.
(3) They identify certain frequencies on the response curve.
(4) They eliminate the back voltage generated in the yoke during retrace.

15. The ripple frequency of a full-wave rectifier circuit in a 60 Hertz power system would be

(1) 60 Hz.
(2) 120 Hz.
(3) 400 Hz.
(4) 800 Hz.

16. An oscilloscope is calibrated for a ten volt-per-inch deflection. If a 5-volt (rms) sine-wave voltage is fed to the vertical terminals of the scope, the deflection should be

(1) exactly 1.0 inch.
(2) about 2.8 inches.
(3) about 1.11 inches.

(4) about 1.4 inches.

17. If the video signal is applied to the grid of the picture tube, the sync pulses

(1) will be the most negative part of the signal.
(2) will be the most positive part of the signal.

18. An advantage to using a triggered-sweep for viewing the color burst is that

(1) it shows the color lobes in a "daisy pattern."
(2) it shows the burst in a different color.
(3) it makes it possible to lock the scope onto the burst.
(4) it has a higher frequency response.

19. The color burst is located

(1) on the front porch of the vertical sync pulse.
(2) on the back porch of the horizontal sync pulse.
(3) on the back porch of the vertical sync pulse.
(4) on the front porch of the horizontal sync pulse.

20. The horizontal deflection coils are located

(1) at the right and left sides of the yoke.
(2) only at the top side of the yoke.
(3) only at the left side of the yoke.
(4) at the top and bottom of the yoke.

21. If the petals of the "daisy" on the vector scope are rotating,

(1) the 3.58-MHz oscillator is not synchronized.
(2) the sync separator is not working properly.

22. To align the chroma bandpass amplifiers, use

(1) a vectorscope and a color-bar generator.
(2) a triggered sweep and a curve tracer.
(3) a sweep generator and an oscilloscope.
(4) an accurate rf generator and a VTVM.

23. If a pure sine wave is applied to an integrating circuit, the output will be

(1) pulses.
(2) a square wave.
(3) a triangular wave.
(4) a sine wave.

24. Which of the following may occur if the video amplifier stage in a color TV receiver has a poor low-frequency response?

(1) The vertical blanking bar will not have a constant shade of grade from start to finish.
(2) The sound will be very weak.

25. Vertical sync pulses

(1) come from the flyback transformer.
(2) come from the AGC amplifier.
(3) come directly from the sync separator.
(4) come from an integrator.

26. A transistor junction may be damaged by excessive current flow when an ohmmeter is used to measure the forward resistance, if the ohmmeter is in the

(1) R × 1 position.
(2) R × 10 k position.

27. When soldering parts into radios, use

(1) 60-40 solder.
(2) 40-60 solder.

28. The collector of a power transistor may be connected

(1) through the smallest terminal lead.
(2) through the case or through a threaded stud.

29. A true square wave would be obtained by

(1) combining a fundamental-frequency sine wave with an infinite number of odd harmonic sine waves, each having the proper amplitude and phase.
(2) combining a fundamental-frequency sine wave with an infinited number of even harmonic sine waves, each having the proper amplitude and phase.

30. The waveform of the voltage delivered to the deflection yoke is a sawtooth sitting on a square pedestal. The reason for using this waveform instead of using a simple sawtooth current is

(1) to overcome inductive kickback in the yoke.
(2) to decrease the amount of heating that will take place in the yoke.
(3) to produce the necessary ringing in the yoke during retrace.
(4) to increase the rms value of the waveform.

31. The color killer cuts off the

(1) video amplifier during color reception.
(2) bandpass amplifier during monochrome reception.

32. Wrong colors are not likely to be caused by

(1) an incorrectly adjusted contrast control.
(2) improper adjustment of the hue control.

33. Diagonal stripes or variegated colors are most likely indicated by

(1) incorrect receiver agc voltage.
(2) a defective video amplifier stage.
(3) a defective rf amplifier.
(4) a loss of color synchronization.

34. White on the screen of a color picture tube is obtained when

(1) all three guns are conducting.
(2) the color-burst signal is at its maximum amplitude.

35. Which of the following is the lower value?

(1) The rms value of a sine-wave voltage.
(2) The average value of a sine-wave voltage.

36. A sine-wave voltage is measured with an oscilloscope and found to be 15 volts peak-to-peak. What is the rms value of this voltage?

(1) 10.6 V
(2) 9.54 V
(3) 7.2 V
(4) 5.3 V

37. A possible cause of green and magenta bars appearing on the screen is

(1) hum modulating the blue and red guns.
(2) hum modulating the green gun.
(3) an open green gun.
(4) an open video detector.

38. When signals from the color demodulators do not have sufficient amplitude, the result is

(1) loss of brightness.
(2) desaturated colors.
(3) loss of gray scale.
(4) improper hues.

39. To get a Lissajous pattern on an oscilloscope

(1) feed a sine-wave voltage to the vertical deflection plates and a sawtooth wave to the horizontal deflection plates.
(2) feed a sine-wave voltage to both the vertical and the horizontal deflection plates.

(3) feed a sine-wave voltage to the horizontal deflection plates and a sawtooth wave to the vertical deflection plates.

40. An oscilloscope will display a circle when

(1) two sine waves that are 90° out of phase are fed to the vertical and horizontal deflection plates.
(2) two in-phase sine waves are fed to the horizontal and the vertical deflection plates.

41. In order to align the color section, the color-bar generator output terminals are connected to the

(1) frequency control for the color subcarrier generator.
(2) luminance amplifier after the delay line.
(3) color picture tube.
(4) antenna terminals.

42. As a first step in troubleshooting a color television receiver

(1) determine if the horizontal and vertical hold controls are working.
(2) determine if the contrast control will operate throughout its range.
(3) measure the high voltage.
(4) determine if the trouble is in the monochrome or color section.

43. The eye is most sensitive to the

(1) red wavelength.
(2) green-yellow wavelength.
(3) blue wavelength.
(4) ultraviolet end of the spectrum.

44. Which of the following has the greatest bandwidth in the composite television signal?

(1) The I signal.
(2) The audio signal.
(3) The luminance signal
(4) The Q signal

45. The current drawn by a transistor radio is measured by connecting a milliammeter across the ON-OFF switch. Which of the following is true?

(1) The switch must be in the ON position.
(2) An ohmmeter should be connected across the switch.
(3) If the radio has a Class-B audio amplifier, the current drain

will decrease when the radio is tuned to a station.

(4) If the radio has a Class-B audio amplifier, the current drawn will increase when the radio is tuned to a station.

46. The forward emitter-base resistance on a certain transistor is 220 ohms, and the reverse resistance is 500 ohms. Which of the following is correct?

(1) The ratio is too high, the transistor is not good.
(2) The ratio is too low; the transistor is not good.

47. The base of an NPN transistor is a Class-A amplifier circuit has a voltage of −6.8 V. Which of the following would you expect to measure on the emitter?

(1) +3 V
(2) +7.2 V
(3) −7.2 V
(4) −3 V

48. An ohmmeter is connected across the emitter-collector terminals of a transistor. When the collector and base leads are shorted together, the ohmmeter shows a lower resistance. Which of the following is true?

(1) The transistor is shorted.
(2) This is normal for a good transistor.
(3) The emitter-base junction of the transistor is open.
(4) This can never happen.

49. The collector circuit of a certain PNP transistor is connected to ground through a transformer winding. You would expect the emitter of this transistor to be

(1) positive with respect to ground.
(2) negative with respect to ground.

50. To look at the video amplifier signal with a scope, you should use a

(1) low-capacity probe.
(2) voltage-doubler probe.
(3) demodulator probe.
(4) direct probe.

51. The bias on a certain transistor measures too low, but its collector current is too high. A likely cause is

(1) an open voltage divider circuit for biasing the base.
(2) a leaky transistor.

(3) an open bypass capacitor in the emitter stage.
(4) an open emitter resistor.

52. A certain transistor is not conducting. The base bias voltage is measured and found to be OK. A possible cause is

(1) the power supply is not operating.
(2) the transistor is open.
(3) the transistor is shorted.
(4) the power supply is at fault.

ANSWERS TO PRACTICE TEST

Question Number	Answer Number
01	(2)
02	(2)
03	(2)
04	(2)
05	(2)
06	(4)
07	(1)
08	(1)
09	(1)
10	(2)
11	(3)
12	(4)
13	(1)
14	(3)
15	(2)
16	(4)
17	(1)
18	(3)
19	(2)
20	(4)
21	(1)
22	(3)
23	(4)
24	(1)
25	(4)
26	(1)
27	(1)
28	(2)
29	(1)
30	(1)
31	(2)
32	(1)
33	(4)
34	(1)
35	(2)
36	(4)
37	(2)
38	(2)
39	(2)

Question Number	Answer Number
40	(1)
41	(4)
42	(4)
43	(2)
44	(3)
45	(4)
46	(2)
47	(3)
48	(2)
49	(1)
50	(1)
51	(2)
52	(2)

Appendix

Appendix
A Sample Journeyman CET Test

One problem technicians have with the CET test is that they do not know how to pace themselves for taking a test that may require as long as one or two hours to complete. If it has been a long time since they took a test of that length, they are likely to get tired toward the end. That leads to hurrying, and the result is carelessly missing questions.

You should take this practice test at one setting. Don't even start it unless you know you can take time to go all the way through. Time yourself to get an estimate of how long you will need to take for the actual test.

Grade your paper using the answer sheet at the end of the test. If you miss more than 25% of the questions in any one section you should spend some time reviewing the related subject matter. The subject headings in this practice test are identical to those in the actual Journeyman-Level Consumer Test. Remember that questions 1-75 are in the Associate-level Test. That test has sections 1 through 8.

SECTION 9, ANTENNAS AND TRANSMISSION LINES

76. Which of the antennas shown in Fig. A-1 is especially well suited for eliminating ghosts, but has the disadvantage of being suitable for use in only 1 or 2 channels?

(1) the one shown in A.
(2) the one shown in B.
(3) the one shown in C.

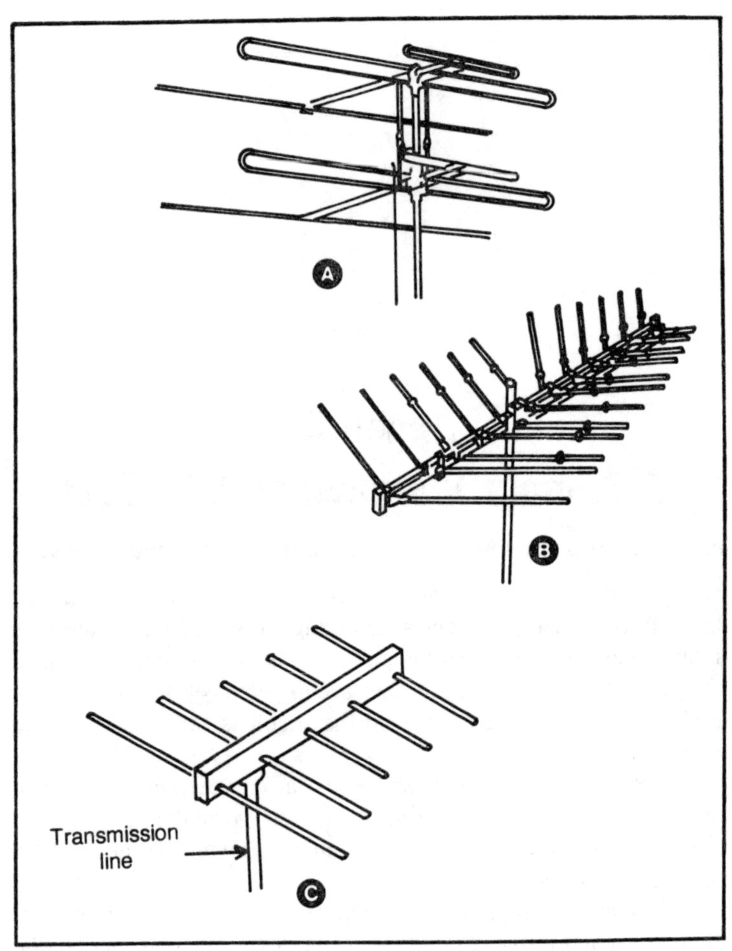

Fig. A-1.

77. Disregarding any change in impedance that may occur at the junction, when you connect two 50-foot lengths of 300-ohm transmission line together (end-to-end) the impedance of the resulting 100-foot line is

(1) 50-ohms.
(2) 75-ohms.
(3) 200-ohms.
(4) 300-ohms.
(5) 500-ohms.

78. The ideal standing-wave ratio on a television transmission line would be

(1) 100:1.
(2) 1:1.
(3) 1:0.
(4) 0:1.
(5) 0:0.

79. The chance of stray electromagnetic pickup is minimized by

(1) running two 300-ohm transmission lines in parallel.
(2) the use of a matching stub.
(3) the use of a Q strap.
(4) the use of coaxial cable.
(5) increasing the length of the horizontal run for the transmission line.

80. If you are looking at a ghost caused by multipath distortion, it

(1) should be tunable with the fine tuning control.
(2) should not be seriously affected by the fine tuning control.

SECTION 10, DIGITAL CIRCUITS IN CONSUMER PRODUCTS

81. Figure A-2 shows the symbol for

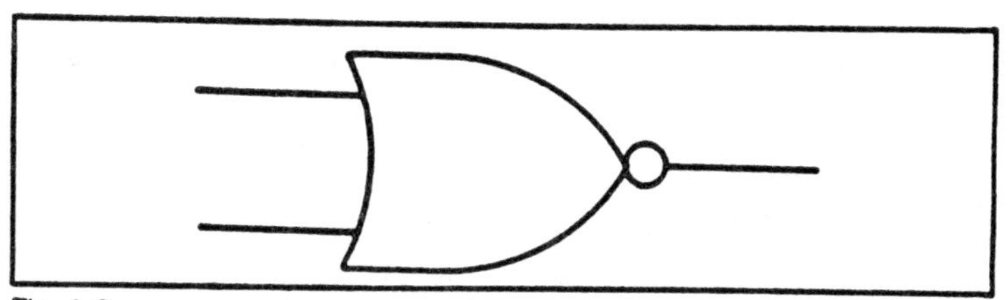

Fig. A-2.

(1) an OR gate.
(2) a NOR gate.
(3) an EXCLUSIVE OR gate.
(4) an AND gate.
(5) an inverter.

82. The arrow in Fig. A-3 points to

Fig. A-3.

(1) pin 7.
(2) pin 12.

83. Is this statement correct: if you measure the logic levels shown on the gate in Fig. A-4 it must be working properly.

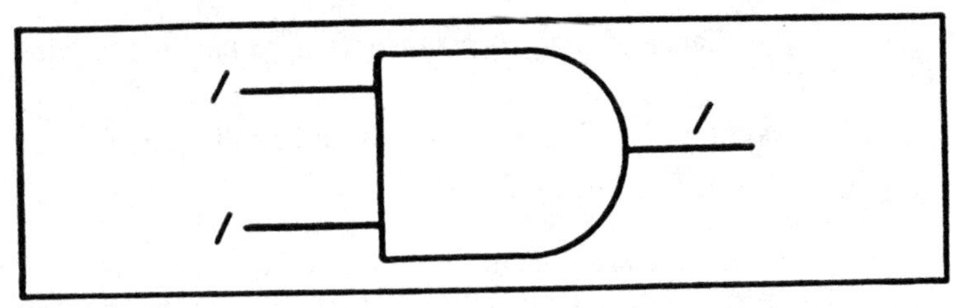

Fig. A-4.

(1) The statement is correct.
(2) The statement is not correct.

84. Which of the following might be used for delaying a signal?

(1) Flip-flop.
(2) Diplexer.
(3) Duplexer.
(4) Multiplexer.
(5) Bucket Brigade.

85. For the input signals shown to the gate in Fig. A-5 the output should be

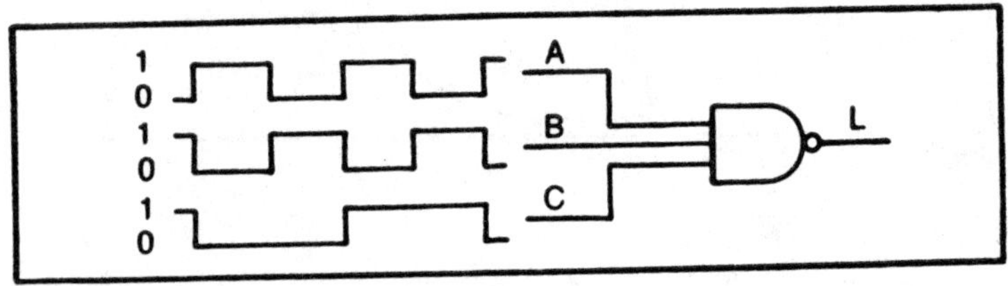

Fig. A-5.

(1) a pulse.
(2) a square wave.
(3) a logic 0 at all times.
(4) a logic 1 at all times.
(5) a sawtooth waveform.

86. The input marked "C" on the gate in Fig. A-5 is permanently connected to common. Now the output will be

(1) a pulse.
(2) a square wave.

(3) a logic 0 at all times.
(4) a logic 1 at all times.
(5) a sawtooth waveform.

87. Which of the symbols in Fig. A-6 represents a NAND gate?

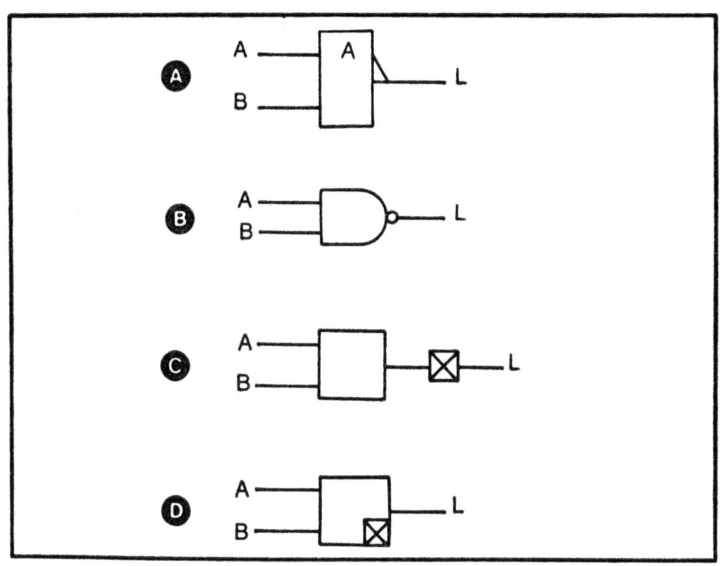

Fig. A-6.

(1) Only the one marked A.
(2) Only the one marked B.
(3) Only the one marked C.
(4) Only the one marked D.
(5) All are symbols used to represent NAND gates.

88. Figure A-7 shows a relay logic circuit for

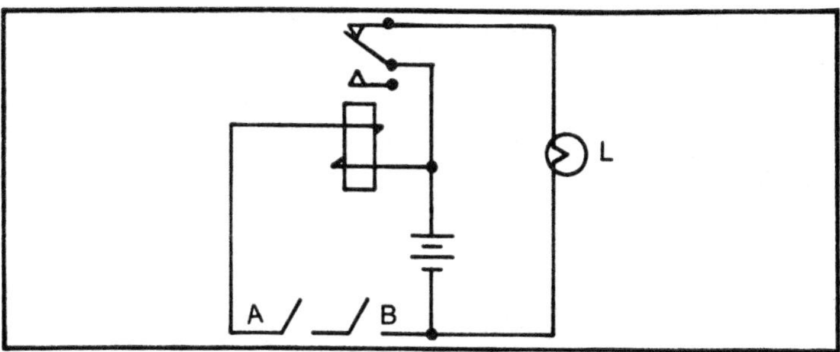

Fig. A-7.

(1) a NAND.
(2) an OR.
(3) an EXCLUSIVE OR.
(4) a NOR.
(5) an AND.

89. The truth table in Fig. A-8 is for

A	B	C	L
0	0	0	0
0	0	1	1
0	1	0	1
0	1	1	1
1	0	0	1
1	0	1	1
1	1	0	1
1	1	1	1

Fig. A-8.

(1) a NAND.
(2) an OR.
(3) an EXCLUSIVE OR.
(4) a NOR.
(5) an AND.

90. Is this equation correct: $\overline{A} + \overline{B} = \overline{AB}$?

(1) It is correct.
(2) It is not correct.

91. Which of the following logic families is normally the fastest?

(1) TTL.
(2) CMOS.
(3) ECL.
(4) RTL.
(5) DTL.

92. Which of the following logic families operates with a −5 V supply?

(1) TTL.

(2) CMOS.
(3) ECL.
(4) RTL.
(5) DTL.

93. For the logic probe in Fig. A-9 the LED marked 'X' will be on when the probe tip touches

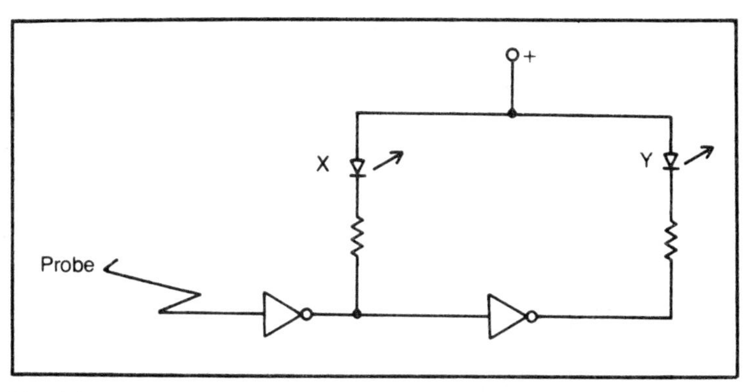

Fig. A-9.

(1) logic 0.
(2) logic 1.

94. To make a RAM, the manufacturer may use

(1) fuses.
(2) flip-flops.
(3) decoders.
(4) encoders.
(5) none of these choices is correct.

95. What is the decimal equivalent of binary 1101?

(1) 10.
(2) 11.
(3) 12.
(4) 13.
(5) 14.

SECTION 11, LINEAR CIRCUITS IN CONSUMER PRODUCTS

96. Which of the following amplifier configurations is best for high-frequency operation?

(1) Common emitter.
(2) Common base.
(3) Common collector.

97. Which of the following is the same as beta?

(1) h_{FB}.
(2) h_{FC}.
(3) h_{FE}.
(4) h_{FR}.
(5) None of these.

98. Whenever you increase the gain of an amplifier you automatically.

(1) increase the bandwidth.
(2) decrease the bandwidth.

99. In the phase-locked loop in Fig. A-10 the block marked 'X' is

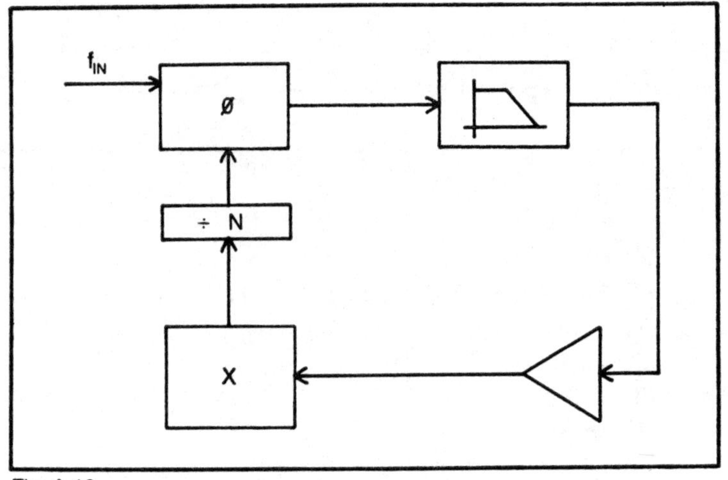

Fig. A-10.

(1) a flip-flop.
(2) a counter.
(3) a decoder.
(4) an operational amplifier.
(5) a VCO.

100. In the operational amplifier circuit in Fig. A-11, the maximum gain occurs when switch SW is in position

(1) X.
(2) Y.
(3) Z.

101. The input stage to an integrated circuit operational amplifier is

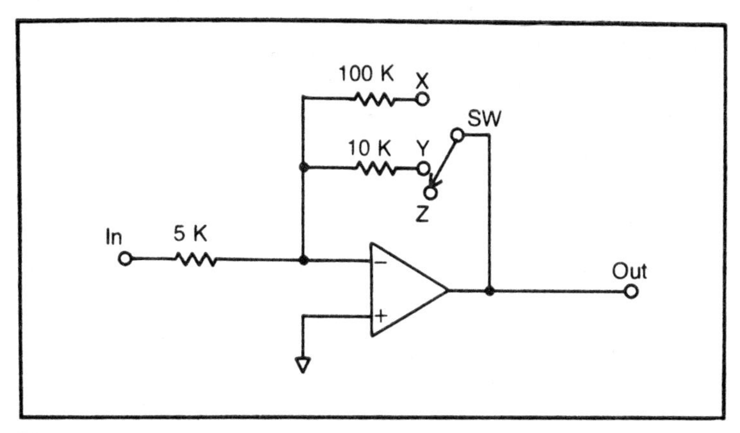

Fig. A-11

(1) a totem pole.
(2) a two-input NOR.
(3) a differential amplifier.
(4) a Darlington Pair.
(5) an R-S flip-flop.

102. To connect an operational amplifier in a common mode configuration

(1) ground the output.
(2) connect the output to the inverting input terminal.
(3) connect the output to the non-inverting input terminal.
(4) connect the inverting and non-inverting terminals together.
(5) connect the two input terminals and the output terminal to B+.

103. The circuit in Fig. A-12 is

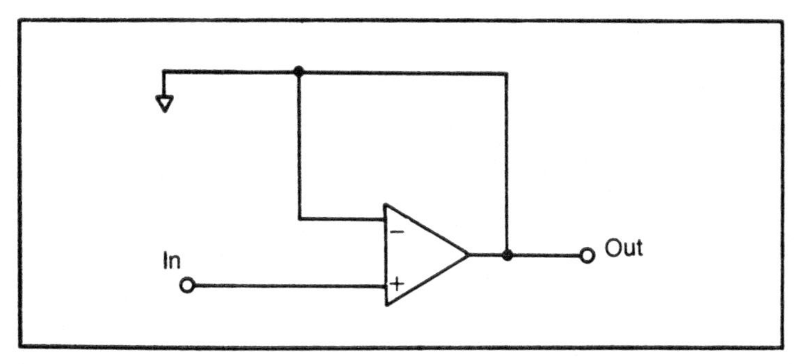

Fig. A-12.

(1) useless.
(2) surely going to be destroyed.
(3) an oscillator.
(4) a buffer.
(5) an amplifier with a very high gain.

104. The battery in the parallel-tuned circuit of Fig. A-13 is

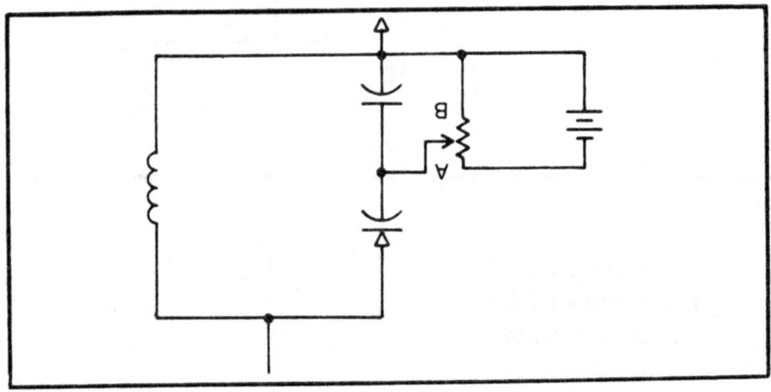

Fig. A-13.

(1) unnecessary.
(2) connected backwards.
(3) properly connected.

105. Moving the arm of the variable resistor toward 'a' in the circuit of Fig. A-13 will

(1) not affect circuit resonance because the battery is connected backwards.
(2) increase the resonant frequency of the circuit.
(3) decrease the resonant frequency of the circuit.

106. To decrease the high-frequency response of the circuit in Fig. A-14 you should move the arm of the variable resistor toward

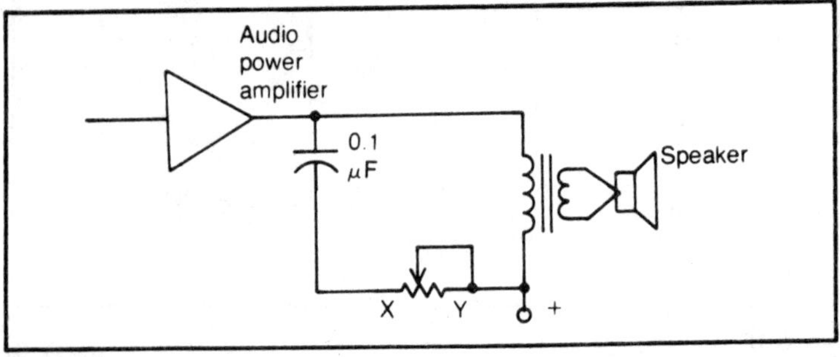

Fig. A-14.

(1) point X.
(2) point Y.

107. You should expect to see an operational amplifier listed under

(1) analog integrated circuits.
(2) thick film circuits.
(3) thin film circuits.
(4) digital integrated circuits.
(5) printed circuits.

108. Which of the following is most likely to produce level shifting?

(1) An NPN/PNP complementary direct coupled pair.
(2) RC coupled amplifier.
(3) Direct-coupled NPN transistors.
(4) An NPN/PNP totem pole configuration with long-tail bias.
(5) Transformer-coupled audio amplifiers.

109. Which of the following is a method of rating differential amplifiers?

(1) Beta squared.
(2) Common mode rejection ratio.
(3) Capture ratio.
(4) Roll off.
(5) Intrinsic standoff ratio.

110. A disadvantage of Darlington amplifiers in power circuits is

(1) low beta.
(2) poor transient response.
(3) poor low-frequency response.
(4) high characteristic noise.
(5) relatively high internal current.

111. A constant-current, 2-terminal component can be made with

(1) a JFET.
(2) a metal film capacitor.
(3) a self-saturating step-up transformer.
(4) a zener diode.
(5) (none of these choices are correct).

SECTION 12, TELEVISION

112. Which of the following is a circuit used to keep the color

oscillator on frequency?

(1) AFC
(2) AFT
(3) AGC
(4) AFPC
(5) ALU

113. Some television receivers automatically adjust the tint using the

(1) VITS signal.
(2) VIRS signal.

114. Which of the following is used to regulate the high-voltage in a color receiver?

(1) VDR
(2) Zener diode
(3) Theristor
(4) Negative feedback fo the horizontal output stage
(5) A combination VDR and thermistor circuit

115. Colored snow during a B&W program might indicate

(1) a defective low-voltage regulator.
(2) the ABL circuit is not working properly.
(3) the color killer isn't working properly.
(4) a defective antenna.
(5) (none of these choices is correct).

116. Which of the following is likely to be used as a detector in a modular television receiver?

(1) Slope detector
(2) Product detector
(3) Envelope detector
(4) Taylor demodulator
(5) None of these answers is correct.

117. To operate the keyed AGC circuit it is necessary to have a composite video input and

(1) a horizontal pulse from the AFC circuit.
(2) a pulse from the flyback.
(3) an AGC input.
(4) a signal from the AFT circuit.
(5) an input from the ABL.

118. You would expect to find a comb filter

(1) in the audio section.

(2) used to separate video and color signals.
(3) in the tuner.
(4) in the vertical oscillator circuit.
(5) in the low voltage power supply.

119. Peaking compensation would most likely be found in

(1) the tuner.
(2) the power supply.
(3) in the Y amplifier.
(4) in the sound section.
(5) in the high-voltage section.

120. A crowbar is

(1) an overvoltage protection circuit.
(2) used to separate sound and picture signals.
(3) a burglar tool—not an electronic circuit.
(4) used to separate H and V signals.
(5) none of these choices is correct.

121. To synchronize a relaxation oscillator, the free-running frequency of the oscillator should be

(1) higher than the sync signal.
(2) lower than the sync signal.

122. The brightness control is most likely to be located in

(1) the high-voltage rectifier.
(2) the horizontal output stage.
(3) the horizontal oscillator.
(4) the direct-coupled video amplifier stage.
(5) the low-voltage power supply.

123. The sound takeoff point in a color television receiver is

(1) before the video detector.
(2) at the video detector.
(3) after the video detector.

124. The sound i-f frequency in a television receiver is

(1) 41.75 MHz.
(2) 45.25 MHz.
(3) 27.5 MHz.
(4) 4.5 MHz.
(5) 120,000 Hertz.

125. A phase-locked loop might be used in

(1) the television tuner.
(2) the low-voltage regulator.

(3) the high-voltage regulator.
(4) the ACL circuit.
(5) the AGC circuit.

126. When you are facing the television receiver the picture is scanned

(1) from left to right.
(2) from right to left.

SECTION 13, VIDEOCASSETTE RECORDERS

127. Which of the following is a disadvantage of a video disc system?

(1) higher cost than video tape systems
(2) cannot be used to record

128. Which of the following is used to read signal information from a video disc?

(1) sonar
(2) radar
(3) maser
(4) laser
(5) eliptical stylus

129. In a video tape recorder the video signal is recorded as

(1) AM
(2) FM

130. To maintain a constant tape speed, a VTR system uses

(1) a cap screw.
(2) a governor on the motor.
(3) an accelerometer.
(4) any of these may be used.
(5) (none of these answers is correct).

131. Which of the following is not a popular recording system?

(1) VHS
(2) cartridgevision
(3) Betamax

SECTION 14, TROUBLESHOOTING CONSUMER EQUIPMENT

132. The waveform in Fig. A-15 has been obtained with a square-wave test. It shows

(1) a loss of low frequencies.
(2) a loss of high frequencies.

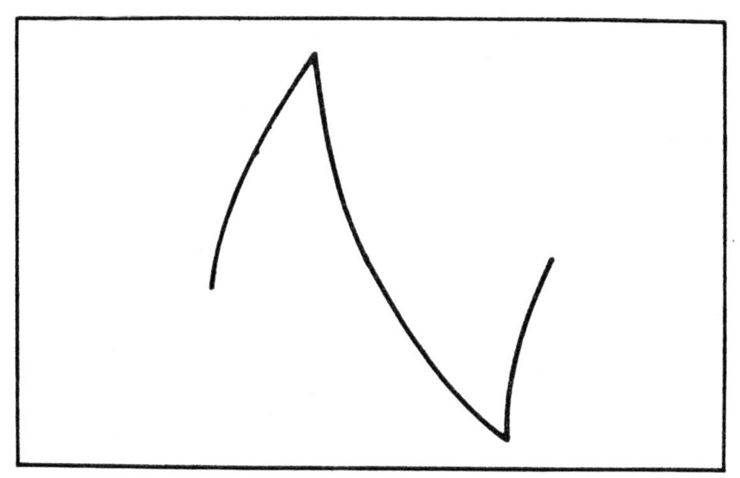

Fig. A-15

133. To check Q, in the circuit of Fig. A-16, a technician shorts its emitter to its base and measures the collector voltage. This test

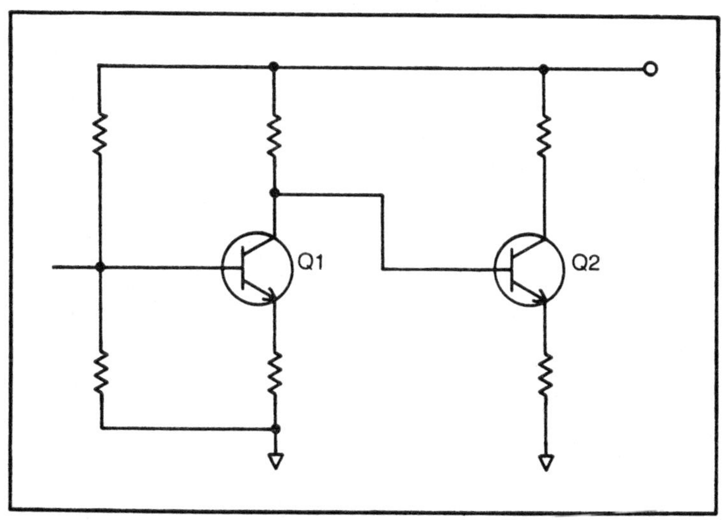

Fig. A-16

(1) is the best method of determining the condition of Q1.
(2) will destroy Q1.
(3) will destroy Q2.

134. You would expect to find horizontal synchronizing pulses at the output of the

(1) differentiator.
(2) integrator.

135. In a color television receiver you would expect the color circuitry to be off when

(1) the color killer is ON.
(2) the color killer is OFF.

136. The time for one line of sweep in a television receiver is about

(1) 123 microseconds.
(2) 94.3 microseconds.
(3) 63.4 milliseconds.
(4) 56.5 milliseconds.
(5) (none of these choices is correct).

137. In which of the following sections would you expect to find the automatic degaussing circuit?

(1) the video amplifier
(2) the high-voltage section
(3) the low-voltage section
(4) the horizontal sweep section
(5) the tuner

138. You might find an LDR in

(1) the vertical sweep circuit.
(2) the horizontal sweep circuit.
(3) the low drain resistor circuit.
(4) the brightness circuit.
(5) the high-voltage supply.

139. The ACC circuit is not working properly. This will cause improper gain in the

(1) bandpass amplifier.
(2) color sync circuit.
(3) video amplifier.
(4) tuner.
(5) (none of these choices is correct).

140. The VITS signal is located

(1) on the horizontal blanking pedestal.
(2) on the vertical blanking pedestal.

141. The AGC voltage in a transistorized television receiver

(1) should be a negative dc.
(2) should be a positive dc.

(3) could be either a positive or negative dc.
(4) may be a square wave.
(5) should be a sinusoidal ac.

142. Which of the following requires a startup circuit?

(1) scan-derived supply
(2) analog regulator
(3) high-voltage regulator
(4) phase-locked oscillator
(5) AFT.

143. If a surge-limiting resistor is burned out

(1) replace it with one having a higher power rating; but, never replace it with one having a lower power rating.
(2) expect the high-voltage measurement to be high.
(3) there will be no dc output from the low-voltage supply.
(4) the set must have been struck by lightning.
(5) it will destroy the audio power transistor.

144. Use a ringing test to check

(1) continuity in the high-voltage section.
(2) the yoke.
(3) the video amplifiers.
(4) start the horizontal oscillator.
(5) AGC recovery time.

145. The i-f frequency of a broadcast FM receiver is

(1) 4.5 MHz.
(2) 10.7 MHz.
(3) 455 kHz.
(4) 41.25 kHz.
(5) (none of these choices is correct).

146. When using a Lissajous pattern to check an audio amplifier for distortion you should get

(1) an ellipse.
(2) a circle.
(3) a square.
(4) a straight line.
(5) a pattern of dots.

SECTION 15, TEST EQUIPMENT

147. Which of the following is a method of evaluating transistors in a common base configuration?

(1) Alpha
(2) Beta
(3) Gamma
(4) Slewing rate
(5) CMRR

148. Which of the following is used to convert a sensitive meter movement to a voltmeter?

(1) Shunt
(2) Multiplier

149. A meter movement with a full-scale deflection reading of 10 microamperes is used as a voltmeter. The meter has a rating of

(1) 10,000 ohms per volt.
(2) 20,000 ohms per volt.
(3) 50,000 ohms per volt.
(4) 100,000 ohms per volt.
(5) (the answer cannot be determined from the information given).

150. Which of the following would be useful for looking at the VITS signal?

(1) A 5-megahertz recurrent-sweep oscilloscope
(2) A prescaler
(3) A logic analyzer
(4) A vector scope
(5) A scope with a delayed sweep

ANSWERS TO PRACTICE CET TEST

Question Number	Answer Number
76	(3)
77	(4)
78	(2)
79	(4)
80	(2)
81	(2)
82	(1)
83	(2)
84	(5)
85	(4)
86	(4)
87	(5)
88	(1)
89	(2)
90	(1)
91	(3)
92	(3)
93	(2)
94	(2)
95	(4)
96	(2)
97	(3)
98	(2)
99	(5)
100	(3)
101	(3)
102	(4)
103	(4)
104	(3)
105	(2)
106	(1)
107	(1)
108	(3)
109	(2)
110	(5)
111	(1)
112	(4)
113	(2)
114	(4)

Question Number	Answer Number
115	(3)
116	(5)
117	(2)
118	(2)
119	(3)
120	(1)
121	(2)
122	(4)
123	(1)
124	(4)
125	(1)
126	(1)
127	(2)
128	(4)
129	(2)
130	(5)
131	(2)
132	(2)
133	(3)
134	(1)
135	(1)
136	(5)
137	(3)
138	(4)
139	(1)
140	(2)
141	(3)
142	(1)
143	(3)
144	(2)
145	(2)
146	(4)
147	(1)
148	(2)
149	(4)
150	(5)

Index

Index

A

A/D converters, 131
Address bus, 120
Agitation noise, 16
Alpha cutoff, 60
ALU, 120
Aluminum wire, 89
Amplification, classes of, 51
Amplifier, cascade, 55
Amplifier, class-A, 59
Amplifier, differential, 155
Amplifier, inverting, 157
Amplifier, noninverting, 158
Amplifier configurations, 56, 57
Amplifier coupling circuits, 164
Amplifiers, distortion in, 59
Amplifiers, noise in, 55
Amplifiers, testing, 229
Analog circuits, 151
Analog computers, 152
AND, 105
Answer sheets, 6-8
Antenna installations, 80, 87
Antennas, 80, 84
Antennas and transmission lines, 4
Arithmetic logic unit, 120
Asynchronous, 129

B

Balanced line, 80
Balun, 82
Bandpass amplifier, 193
Bandwidth, 59
Base, 112
Beta cutoff, 60
Beta-squared amplifier, 53
Bilateral circuit component, 16
Bit slice, 128
Bode plot, 60, 153
Boolean algebra, 105
Boolean equation, 106
Boolean equations for logic gates, 107
Bootstrap, 53
Bootstrap circuit, 52
Bowtie antenna, 87
Breakover device, 27
Buffer, 159
Buffer amplifier, 160
Buses, 122

C

Capacitor, ceramic, 21
Capacitor, mica, 21
Capacitor, nonpolarized electrolytic, 22
Capacitor, paper, 21
Capacitor, tantalum, 21
Capacitor, variable, 22
Capacitors, 20
Carrier reinsertion, 191
Cascade amplifier, 55

Cascoded, 164
CET, journeyman, 1
CET certificate, 2
CET test, 1
CET test, a sample, 257-276
CET test questions, 243-253
CET test questions, practice, 36-42, 72-79, 98-104, 141-150, 181-187, 207-219
CET test rules, 9, 10
Charge-coupled devices, 131
Chroma module, 199
Circuits, ac, 1
Circuits, dc, 1
Circuits, time constant, 47
Clock, 119
CMOS, 129
CMOS flip-flops, 117
Coaxial cable, 80
Colored confetti, 199
Common base, 52, 229
Common cathode, 52
Common collector, 52
Common drain, 52
Common emitter, 52, 229
Common gate, 52
Common grid, 52
Common source, 52
Compander, 160
Components, electronic, 3
Control bus, 120
Counters, programmable, 130
Counting systems, 111
Crossover distortion, 233
Current limiter, 168

D

D/A converters, 131
Data bus, 120
Decoders, 130
DeMorgan's theorem, 111
Diac, 23
Differential amplifier, 155
Digital circuits, 4, 105
Diode, four-layer, 27
Diode, hot-carrier, 27
Diode, light-activated, 26
Diode, light-emitting, 26
Diode, Schottky, 27
Diode, varactor, 27
Diode, Zener, 27
Diodes, 24
Diodes, constant-current, 25
Diodes, optoelectronic, 26
Diodes, tunnel, 25
Dipole, 84
Direct coupled, 164
Director, 85
Distortion, multi-path, 86
Dolby audio systems, 163
Dynamic memory, 131

E

ECL, 129
Emitter-coupled logic, 129
Emitter follower, 52
Encoders, 130
Exclusive NOR, 105
Exclusive OR, 105
Expander, 160

F

FET amplifier, 52
Filter, active, 161
Filter, comb, 198
Filters, LC, 45
Firing potential, 23
Flicker noise, 58
Flip-flop, J-K, 117
Flip-flop, R-S, 116
Flip-flop, type-D, 118
Flip-flops, 115
Flyback transformer, 196
FM, 162, 192
FM signals, 188
Full-wave doublers, 167

G

Gain, 59
Glitches, 129
Ground rod, testing a, 90

H

Half-wave doublers, 167
Harmonic distortion, 59
Hertz antennas, 84

I

Impedance, characteristic, 81
Impedance, surge, 81
inductors, 22
Instruments, 3
Intermodulation distortion, 59
Inverting amplifier, 157

J

JFET, 25
Johnson noise, 16
Journeyman test options, 4

L

LAD, 26
LED, 26

Level shifting, 165
Lightning, 88
Linear circuit elements, 15
Linear circuits, 5, 151
Linear rolloff, 153
Lissajous pattern, 224
Logic comparator, 105
Logic gate symbols and simple circuits, 108, 109, 110
Log periodic antenna, 86
Long-tailed bias, 54

M
Magnetron, 23
Marconi antenna, 85
Mathematics, basic, 1
Memories, 119
Meter movements, 221
Meter movements, analog, 222
Microprocessors, 119
Microprocessor terminology, 123-126
Mirrored scale, 223
Modulo, 112
MOSFET, 51, 53, 58
Multi-path distortion, 86
Multiplexer, 121

N
NAND, 105
Network analysis, 3
Noninverting amplifier, 158
NOR, 105
NOT, 105
Number systems, 112

O
Ohmmeter test for transistors, 232
Ohm's law, 3, 221
Operational amplifier, 152, 159
Operator, 152
Optical coupler, 26
OR, 105
Oscillator, 164
Oscillator, relaxation, 48
Oscilloscopes, 223

P
Pad, 82
Parallel data, 127
Parasitic elements, 85
Partition noise, 55
Phase-locked loop, 161
Phaseshift distortion, 59
Pilot, 192
Potentiometers, 19
Power supply, scan-derived, 197
Power supply circuits, 166
Probes, high-voltage, 227
Program counter, 128
Programmed review, Chapter 1, 10-14
Programmed review, Chapter 2, 28-36
Programmed review, Chapter 3, 60-72
Programmed review, Chapter 4, 92-98
Programmed review, Chapter 5, 131-141
Programmed review, Chapter 6, 170-181
Programmed review, Chapter 7, 200-207
Programmed review, Chapter 8, 235-243
PROM, 121

R
RAM, 121
Random access memory, 121
Receiver, color, 195
Receiver, monochrome, 194
Rectifiers, 167
  half-wave
  full-wave
  bridge
Reflector, 85
Reflex circuits, 57
Register, 127
Regulator, switching, 169
Regulators, tracking, 164
Resistor, carbon-composition, 17
Resistor, carbon-film, 18
Resistor, metal-film, 18
Resistor, wire-wound, 19
Resistor-capacitor coupling, 164
Resistor color code, 17
Resistors, 16
Rheostats, 19
Ripple, 165
ROM, 121

S
Sample and hold, 161
SCR, 169
Semiconductors, 3
Serial data, 127
Series multiplier, 223
Shot noise, 58
Shunt resistor, 223
6802 pinout, 125
Slewing rate, 156
Standing waves, 83
Static memory, 131
Synchronous, 129

Sync pulse, 190

T

Television, 5, 188
Television receiver, 193
Test equipment, 5, 220
Tests and measurements, 3
Thermistors, 27
Three-state buffer, 113
Three-state inverter, 114
Three-terminal amplifying components, 44
Timers, 160
Transformer, flyback, 196
Transmission lines, 80
Transmission lines, installation of, 83
Transistors, 3
Troubleshooting, 3, 5, 220
Truth table, 105
Truth tables for basic gates, 106
TTL, 129
TTL flip-flops, 117
Tubular twin lead, 82
TV, color, 192
TV channel signals, 189
Twin lead, 80
Two-terminal components, 15
Two-terminal components, nonlinear, 23

U

Unbalanced line, 81

V

Varistors, 27
VCO, 162
VDR, 27
Videocassette recorders, 5
Video recorders, 199
VIRS, 191

W

Waveforms, 228

Edited by Roland S. Phelps

Other Bestsellers From TAB

☐ **BASIC INTEGRATED CIRCUITS—Marks**

With the step-by-step guidance provided by electronics design expert Myles Marks, anyone with basic electronics know-how can design and build almost any type of modern IC device for almost any application imaginable! In fact, this is the ideal sourcebook for every hobbyist or experimenter who want to expand their digital electronics knowledge and begin exploring the fascinating new possibilities offered by today's IC technology. 432 pp., 319 illus.

Paper $16.95 Hard $26.95
Book No. 2609

☐ **THE ENCYCLOPEDIA OF ELECTRONIC CIRCUITS—Graf**

Here's the electronics hobbyist's and technician's dream treasury of analog and digital circuits—nearly 100 circuit categories . . . over 1,200 individual circuits designed for long-lasting applications potential. Adding even more to the value of this resource in the exhaustively thorough index which gives you instant access to exactly the circuits you need each and every time! 768 pp., 1,762 illus. 7″ × 10″.

Paper $29.95 Book No. 1938

☐ **FUNDAMENTALS OF DIRECT CURRENT**

An essential guide for anyone who wants to learn the basics of electricity and electronics. DC concepts are the easiest to grasp, making them the logical starting point for all electronics practice. This invaluable guide explains all of these basic concepts and gives you the strong foundation you need for more advanced work in electronics. Plus, the appendix contains indispensable DC formulas, electrical terms, abbreviations, and electronic symbols. 252 pp., 164 illus.

Paper $12.95 Hard $19.95
Book No. 1870

☐ **ENCYCLOPEDIA OF ELECTRONICS**

Here are more than 3,000 complete articles covering many more thousands of electronics terms and applications. A must-have resource for anyone involved in any area of electronics or communications practice. From basic electronics terms to state-of-the-art digital electronics theory and applications . . . from microcomputers and laser technology to amateur radio and satellite TV, you can count on finding the information you need! 1,024 pp., 1,300 illus. 8 1/2″ × 11″.

Hard $60.00 Book No. 2000

☐ **POCKET DIGITAL MULTIMETER TECHNIQUES—Davidson**

If there is any single test instrument that belongs on every hobbyist's and technician's workbench, it is the handy little DMM. Not only does it offer portability, but also greater accuracy than older types of multimeters. With it—and the hands-on guidance provided by this sourcebook—you can troubleshoot everything from small electronic components to complex cassette players and TV sets. 320 pp., 390 illus.

Paper $14.95 Hard $22.95
Book No. 1887

☐ **BASIC ELECTRONICS THEORY—WITH PROJECTS AND EXPERIMENTS—2nd Ed.**

If you're looking for an easy-to-follow introduction to modern electronics . . . or if you're an experienced hobbyist or technician in need of a quick-reference guide . . . there's simply no better sourcebook than this guide. It includes all the basics plus the most recent digital developments and troubleshooting techniques. All new material covers AM, stereo, new video sources, computers, and more. 672 pp., 645 illus.

Paper $18.95 Hard $29.95
Book No. 1775